BLUE ECONOMY

This book presents state-of-the-art perspectives on the Blue Economy. It applies important geographical and sustainability transitions perspectives and underscores how Blue Economy dynamics are situated in regional contexts and shaped by the people who live there.

The book highlights the Blue Economy concept as a potential driver of regionally sensitive, ecologically embedded, and community-focused sustainability. The scope for Blue Economy to form a core "cog" in our low-carbon future is obvious, from the potential for renewable energy production and coastal resilience building to possibilities for sustainable food production and the delivery of economic opportunities for peripheral communities. However, fundamental questions remain on how to meaningfully deliver these promises, such as how to avoid embedding a model of damaging extractivism, as per the terrestrial economy, and how to deliver on the key social sustainability principles of human well-being, equity, and justice when planning and developing blue economies. As the UN Decade of Ocean Science for Sustainable Development opens, this book provides a timely reminder of the richness, diversity, and potential of coastal and marine spaces. It advances geographical and transdisciplinary understandings of the Blue Economy and sets a baseline for continued scholarly engagement with the Blue Economy from a variety of perspectives.

This timely contribution will be of interest to policy makers, academics, industry leaders, decision makers, and stakeholders working in or connected to the Blue Economy Sphere and working in the fields of Economic Geography, Regional Development, Public Policy and Planning, Environmental Studies, and Coastal Zone Management.

C. Patrick Heidkamp is a Professor of Geography in the Department of the Environment, Geography and Marine Sciences at Southern Connecticut State

University, USA, and an affiliate faculty member at the University Centre of the Westfjords in Iceland. During much of the editorial work for this volume, he was a DAAD (German Academic Exchange Service) supported Guest Professor in the Department of Geography at the University of Cologne. He is an economic geographer with current research interests in sustainability transitions in the coastal zone and transdisciplinary engagement with the Blue Economy.

John Edward Morrissey is a Lecturer in Geography at Mary Immaculate College (MIC). His research is focused on issues of sustainable development, particularly sustainability transitions, low-carbon development, and challenges of low-carbon economy for urban and coastal communities. John's work is informed by transdisciplinary approaches with a focus on the spatial and social differentiation of transition processes between communities. John has experience of researching sustainability issues in Ireland, Australia, New Zealand, and the UK.

Celine Germond-Duret is a Lecturer in Environmental Politics and Policy at Lancaster University, UK. Her expertise covers marine policy and the Blue Economy, international climate politics and global inequalities, indigenous peoples, as well as discourse analysis. Her work aims at unravelling the power relations and dominant discourses in development and environmental politics and highlighting their practical implications. Her research notably appeared in *Development and Change; Environment, Development and Sustainability; Marine Policy; Third World Quarterly;* and *Sustainable Development*. She co-edited a special issue on the Blue Economy published in *The Geographical Journal* and co-authored the "Blue Economy" entry for the International Encyclopedia of Geography (Wiley-AAG).

The Dynamics of Economic Space

This series aims to play a leading international role in the development, promulgation and dissemination of new ideas in economic geography. It has as its goal the development of a strong analytical perspective on the processes, problems and policies associated with the dynamics of local and regional economies as they are incorporated into the globalizing world economy. In recognition of the increasing complexity of the world economy, the Commission's interests include: industrial production; business, professional and financial services, and the broader service economy including e-business; corporations, corporate power, enterprise and entrepreneurship; the changing world of work and intensifying economic interconnectedness.

Beyond Free Market
Social Inclusion and Globalization
Edited by Fayyaz Baqir and Sanni Yaya

Culture, Creativity and Economy
Collaborative practices, value creation and spaces of creativity
Edited by Brian J. Hracs, Taylor Brydges, Tina Haisch, Atle Hauge, Johan Jansson and Jenny Sjöholm

Social Protection and Informal Workers in Sub-Saharan Africa
Lived Realities and Associational Experiences from Tanzania and Kenya
Edited by Lone Riisgaard, Winnie Mitullah, and Nina Torm

Economies, Institutions and Territories
Dissecting Nexuses in a Changing World
Edited by Luca Storti, Giulia Urso and Neil Reid

Excessive Inequality and Socio-Economic Progress
Ona Gražina Rakauskienė, Dalia Streimikiene and Lina Volodzkienė

Blue Economy
People and Regions in Transitions
Edited by C. Patrick Heidkamp, John Morrissey and Celine Germond-Duret

BLUE ECONOMY

People and Regions in Transitions

*Edited by C. Patrick Heidkamp,
John Edward Morrissey and
Celine Germond-Duret*

Routledge
Taylor & Francis Group
LONDON AND NEW YORK

Cover image: gettyimages.ca/Tan Dao Duy

First published 2023
by Routledge
4 Park Square, Milton Park, Abingdon, Oxon OX14 4RN

and by Routledge
605 Third Avenue, New York, NY 10158

Routledge is an imprint of the Taylor & Francis Group, an Informa business

British Library Cataloguing-in-Publication Data
A catalogue record for this book is available from the British Library

ISBN: 9781032248158 (hbk)
ISBN: 9781032248172 (pbk)
ISBN: 9781003280248 (ebk)

DOI: 10.4324/9781003280248

Typeset in Bembo
by KnowledgeWorks Global Ltd.

CONTENTS

FIGURES

TABLES

CONTRIBUTORS

Ebru Arıcıoğlu is an author of articles and book chapters on economics. Her research focuses primarily on regional development and international trade. She has published articles and book chapters about regional development, international trade, as well as the relationship between energy consumption and economic growth, the convergence of gross domestic product in the EU. She has been teaching at the university level since 2012. She currently teaches graduate and undergraduate regional development, innovation systems, economic integration, and EU courses in the Department of Economics of Mersin University. She also has been working as researcher and advisor in many national and international projects.

Marie Aschenbrenner is a PhD student and teaching assistant at the Research Unit for Economic Geography at LMU Munich, Germany. She is part of an interdisciplinary DFG research network on "Urban Ethics. Conflicts over good and proper urban living in the 20th and 21st centuries" with a project on Auckland's Hauraki Gulf. Her research interests are (urban) environmental planning, marine spatial planning, ethics of environmental care, and political ecology. She also enjoys working in the field of transformative learning and holds a certificate in Education for Sustainable Development.

Stephen Axon is an Assistant Professor of Sustainability Science with the Department of the Environment, Geography, and Marine Sciences at Southern Connecticut State University. His research focuses on the socio-cultural dimensions of sustainability. Principally, his research investigates the challenges of making everyday life and the communities we live in more sustainable; and the application of interventions and "what works" approaches that facilitate and maintain inclusive, equitable, and participatory sustainable lifestyles. This research aims to advance both conceptual understanding and practical application of enhancing the efficacy, form, and function, of public engagement and collective action as part of community-based sustainability. Dr Axon is the Graduate Program Coordinator for the MS Environmental Studies, the Co-Director of the Connecticut State University System Center for Environmental Literacy and Sustainability Education (CELSE), and Editor for The Northeastern Geographer.

Lauren Brideau worked as a University and Research Assistant in Coastal and Marine Science with the Department of the Environment, Geography and Marine Sciences at Southern Connecticut State University. She is currently a PhD student in Environmental Sciences at the University of Virginia. Brideau's research interests include marine ecology, coastal processes, and ecosystem restoration and function. Brideau is also widely interested in using research applications to address current and future coastal sustainability challenges.

Emma Cross is an Assistant Professor of Coastal and Marine Science with the Department of the Environment, Geography and Marine Sciences at Southern Connecticut State University. Her research focuses on the effects of ocean acidification, warming, and hypoxia on seaweed, marine invertebrates, and vertebrates. Recently, her work has focused on climate change mitigation strategies for the aquaculture industry and the environmental impacts of aquaculture on water quality and biodiversity. Dr Cross uses field analyses, long-term laboratory experiments and museum collections as well as collaborations with the aquaculture industry and non-profit organisations to address topical marine science issues.

Michaela Garland is a PhD Student with the Department of Geography at the University of Connecticut. Michaela's research interests are focused on marine social science, sustainability, social justice, and the Blue Economy. Michaela is currently investigating the Long Island Sound's Blue Plan and the potential inclusion of just sustainabilities as well as the current and potential Blue Economy for the Long Island Sound region.

Annette F. Govindarajan is a molecular ecologist in the Biology Department of the Woods Hole Oceanographic Institution. She is interested in the diversity, ecology, and evolution of marine animals, and has particular expertise in the biology of hydrozoans. Her research program includes developing and applying environmental DNA (eDNA) approaches to characterise animal diversity in midwater and coastal ecosystems.

Celine Germond-Duret is a Lecturer in Environmental Politics and Policy at Lancaster University, UK. Her expertise covers marine policy and the Blue Economy, international climate politics and global inequalities, indigenous peoples, as well as discourse analysis. Her work aims at unravelling the power relations and dominant discourses in development and environmental politics and highlighting their practical implications. Her research notably appeared in *Development and Change; Environment, Development and Sustainability; Marine Policy; Third World Quarterly; and Sustainable Development.* She co-edited a special issue on the Blue Economy published in *The Geographical Journal* and co-authored the "Blue Economy" entry for the International Encyclopedia of Geography (Wiley-AAG).

Samuel Hanes is an Associate Professor and Chair of Anthropology at the University of Maine, USA. He obtained his PhD in Geography from Rutgers University. He is a historical geographer with an interest in environmental management. Most of his work examines how people deal with uncertainty in highly complex human-environmental systems. He has studied topics such as the history of conservation and property rights in the US oyster fishery, the evolution of pest management and pollination in the North American blueberry industry, and rural gentrification in coastal Maine, USA.

C. Patrick Heidkamp is a Professor of Geography with the Department of the Environment, Geography, and Marine Sciences at Southern Connecticut State University, USA, and an affiliate faculty member at the University Centre of the Westfjords in Iceland. During much of the editorial work for this volume, he was a DAAD (German Academic Exchange Service) supported Guest Professor with the Department of Geography at the University of Cologne. He is an economic geographer with research interests in sustainability transitions in the coastal zone and transdisciplinary engagement with the Blue Economy.

Miranda Holland is a graduate student pursuing a degree in Integrated Biological Diversity from Southern Connecticut State University. She is also a guest student at the Woods Hole Oceanographic Institution. She is interested in marine biodiversity, ecology, climate change mitigation strategies, environmental DNA (eDNA), and sustainable food systems. Her graduate thesis investigates how multi-species ocean farming impacts biodiversity through eDNA metabarcoding and video analysis.

Teresa Johnson is an Associate Professor in the School of Marine Sciences at the University of Maine, USA. Dr Johnson grew up on the coast of Maine and received her PhD from Rutgers University. A human ecologist by training, her applied marine social science research explores how individuals, communities, and institutions respond to social-ecological change and seeks to understand and improve participatory approaches to science and management. Her current projects focus on marine fisheries, aquaculture, and renewable energy, with particular interest in advancing more inclusive and impactful transdisciplinary research. She teaches graduate and undergraduate courses in marine policy.

Mehmet Ragıp Kalelioğlu is currently working as an Assistant Professor at the Department of Geography, Gaziantep University, Turkey. His research interests concentrate on the sub-areas of Economic Geography such as retail geography, industrial clustering, innovation, and urban economy. Dr Kalelioğlu took part as a researcher in two prestigious projects on innovation and foreign direct investment in the field of regional development. He has also published project reports and research articles in academic journals. Besides, he has presented various papers in national and international congresses. He is currently the Geography Editor of Gaziantep University Journal of Social Sciences.

Miriah Kelly (she/her) is an Assistant Professor of Environmental Science in the Department of Environment, Geography, and Marine Sciences at Southern Connecticut State University in New Haven, CT. Dr Kelly's teaching and research focuses on the human dimensions of ocean and coastal climate change issues. Dr Kelly works in the area of environmental management and decision-making and studies collaborative and transdisciplinary approaches to addressing complex environmental problems. Dr Kelly is also dedicated to translational environmental sciences, including environmental communication, environmental education, and environmental engagement.

Oliver Klein is managing director of Agrarsoziale Gesellschaft (Agri-Social Association), Göttingen, Germany. Previously, he was a researcher (Post-Doc) and lecturer in economic geography at LMU Munich, University of Greifswald, and University of Vechta. He is interested in rural economies and livelihoods, agri-food systems, sustainability transitions, and the Blue Economy.

Matthias Kokorsch is the Director of the master's program in Coastal Communities and Regional Development at the University Centre of the Westfjords in Ísafjörður, Iceland. Matthias holds a PhD from the University of Iceland in Geography, where much of his research centred on two central themes embedded in the theoretical approaches of Adaptive Co-Management and Resilience: fisheries management and regional development. Having a unique focus on fishing villages and Icelandic coastal communities, much of Matthias's research is based on community resilience, regional development, particularly in sparsely populated regions, structural changes of old-industrial areas, and resource management in combination with aspects of justice and decision-making processes.

Louie Krak is an Executive Policy Specialist for the Pennsylvania Department of Environmental Protections working on climate change, energy, and air policy and regulations. He earned a master's degree in Environmental Studies and a bachelor's degree in Geography both from Southern Connecticut State University.

Richard Le Heron is Emeritus Professor of Geography in the School of Environment at the University of Auckland, a past Chair of the International Geographical Union Research Commission on the Dynamics of Economic Spaces, and a former Vice President (Humanities and Social science) of the Royal Society of New Zealand. His present research explores frontiers of trans-disciplinary knowledge, across science and social sciences. It includes co-imagining and applying socio-ecological knowledge, re-commoning of coastal and ocean domains, the co-governance enactive knowledge under risk and uncertainties, the role of personal positioning, disciplinary norms, and hidden world views and their associated narratives in constraining decision choices and decision-making processes.

Nicolas Lewis is a political-economic geographer at the University of Auckland. He is a member of the Leadership Team of the New Zealand Sustainable Seas National Science Challenge and program leader of its Blue Economy theme. His research explores the challenge of doing economy differently, which lies at the heart of humanity's great local-global challenges of the 21st century. His work seeks to develop methodological and conceptual tools for reconfiguring environment-economy relations and enabling the full range of economic actors to do differently. He has recently commenced a major research program investigating the governmental work of global management consultancies in contemporary extra-state governance.

Clemens Lisdat is a Post-Doc researcher at University of Greifswald, Institute of Geography and Geology, Germany. He is a member of the Interdisciplinary Research Center of the Baltic Sea Region (IFZO). His research is focused on the Blue Economy (especially fisheries), policy mobilities, discourse analysis and assemblage thinking.

Greg Martino is an aquaculture industry professional and co-founder of Cottage City Oysters in Martha's Vineyard, Massachusetts. In aquaculture, he works to use innovative methods to deliver the best ocean products in the most eco-friendly and sustainable way. Martino is interested in protecting, maintaining, and restoring local environments and is particularly motivated to work toward biodiversity maintenance.

Dan Martino is an aquaculture industry professional and co-founder of Cottage City Oysters in Martha's Vineyard, Massachusetts. In aquaculture, he works to use innovative methods to deliver the best ocean products in the most eco-friendly and sustainable way. Martino is interested in protecting, maintaining, and restoring local environments and is particularly motivated to work toward biodiversity maintenance.

Emma McKinley is a Research Fellow at Cardiff University. Her research focuses on understanding the complex relationships between society and the sea, taking account of diverse perceptions, attitudes, and values held by different communities and audiences, and considers how this insight can be used to support effective ocean governance. Emma is the founder of the Marine Social Science Network, a global, interdisciplinary community of marine social science researchers and practitioners and is Chair of the Royal Geographic Society's Coastal and Marine Research Group. She is the Co-Chair of the Marine Social Science Task Group of the UK's Marine Science Coordinating Committee, sits on the International Science Advisory Group for MEOPAR and the IOC-UNESCO's Ocean Literacy Research Community.

John Morrissey is a Lecturer in Geography at Mary Immaculate College (MIC). His research is focused on issues of sustainable development, particularly sustainability transitions, low-carbon development, and challenges of low-carbon economy for urban and coastal communities. John's work is informed by transdisciplinary approaches with a focus on the spatial and social differentiation of transition processes between communities. John has experience of researching sustainability issues in Ireland, Australia, New Zealand, and the UK.

Maeve Rourke is a Graduate Student in the Department of the Environment, Geography and Marine Sciences at Southern Connecticut State University. She is currently an environmental analyst for Connecticut's Department of Energy and Environmental Protection with a focus in coastal management. Maeve is interested in sustainable coastal transitions and in developing the Blue Economy in the state.

Johannes Stein is a master's student in the program Coastal and Marine Management at the University Centre of the Westfjords. His final research project focuses on the Finnafjörður port project.

Christine Tamásy is Professor of Human Geography, University of Greifswald, Institute of Geography and Geology, Germany. She holds a PhD from the University of Hannover and has successfully completed her habilitation at the University of Cologne. Broadly speaking, she is an economic geographer with a special interest in the Baltic Sea Region.

José A. Torres PhD was born in Puerto Rico and he earned his doctorate in Geography from the University of Connecticut in 2015. He is currently a Visiting Assistant Professor with the Department of Geography at Oklahoma State University. His research focuses on the Blue Economy and the socioeconomic impacts of the tourism industry in the Caribbean region, and he also researches sustainable urban redevelopment and sustainable urban tourism practices in Lisbon, Portugal.

Gabriela Triay is a University Assistant in the Department of the Environment, Geography, and Marine Sciences and is currently pursuing a master's degree in Environmental Studies at Southern Connecticut State University. Her research focuses on green ports and the role they may play in a transition towards a just and sustainable Blue Economy.

Ezgi Biçer Uçar Graduated from French-speaking Galatasaray University in 2005, Ezgi BICER UCAR completed her post-graduate degree in Sciences Po Aix-en-Provence, France. During her post as the Coordinator of the Presidency of the Association of the Mediterranean Chambers of Commerce and Industry (ASCAME) at the Istanbul Chamber of Commerce, she played an active role in operating ASCAME more effectively. Working as the Advisor to the President (2011–2015) and Vice Secretary-General (2015–2019) in Mersin Chamber of Commerce and Industry, she continues her professional carrier as Vice-Secretary General of the Mersin Chamber of Shipping.

Gordon Winder is Professor of Economic Geography and Sustainability Research, and Deputy Director of the Department of Geography at LMU Munich. Born in New Zealand, he gained his PhD at the University of Toronto, and worked at the University of Auckland before moving to LMU Munich in 2008.

Nuri Yavan is a Full Professor at the Department of Geography at Ankara University, Turkey. His research interests cover three areas: Economic Geography, particularly industrial and FDI location; Regional Development and Policy, specifically innovation, incentives and institutions; and History and Philosophy of Geography, especially critical analysis of Turkish Geography. He has published 8 books, 21 book chapters, 32 research/project reports, and 30 articles in academic journals. He is the Advisory Editor of the Routledge's Dynamics of Economic Space and Springer's Economic Geography Book Series. He currently serves as a Chair of the IGU Commission on the Dynamics of Economic Spaces.

FOREWORD

The Blue Economy is arguably the ocean buzzword of the last decade. The popularity and use of the concept has accelerated at a rapid pace, matched only by the number of ways in which the term has been employed, defined, and co-opted. At the same time, we are experiencing rapid industrialisation of the world's oceans, often in ways that are deleterious to both human and ocean health. In response, we have seen a small, but growing resistance to the concepts of Blue Economy and blue growth with critics arguing that the terms have been used to drive unsustainable and inequitable ocean development.

Regardless of how well the Blue Economy is defined or accepted, it is clear that the concept is likely to be around for some time yet and the influence of Blue Economy discourses on ocean governance shows no signs of abating. In many ways, debates over what we mean by the Blue Economy have distracted significant attention away from the practical realities of how the Blue Economy is being enacted at local, national, regional, and international scales. This edited volume represents a broad and diverse set of observations about these practical realities.

This collection of works reflects the proceedings of the Coastal Transitions Conference, held at Southern Connecticut State University (SCSU) in November 2020. Despite the challenges of the pandemic, this Conference was able to bring together academic, government, industry, and community stakeholders to discuss the role of the Blue Economy in transitioning to a sustainable and just future. The resulting works contained within this volume provide welcome insights into the role of people and place in the Blue Economy. The contributions contained herein demonstrate that a transition to a sustainable and equitable ocean futures require attention to the social and spatial dimensions of governance. They highlight that broad, normative frameworks like the Blue Economy need to be placed in context and implemented in partnership with local communities and other relevant stakeholders.

I congratulate the editors and conference organisers for bringing together such as breadth of knowledge and experiences. I see this collection of papers as an invitation for further discussion, research, and, importantly, action on ensuring that the Blue Economy is place based, contextually sensitive, and a driver of transformative change.

Dr Michelle Voyer
Snr Research Fellow
Australian National Centre for Ocean Resources and Security (ANCORS)
University of Wollongong, Australia

ACKNOWLEDGEMENTS

The editors would like to sincerely thank the following people for their input and support to the development of this edited volume. From conception to final delivery, it was a combined team effort and the finished product would not have been possible without the enthusiasm and expertise of these key people.

Many thanks to all contributors and authors for their superb research, prompt delivery, and open engagement. We would like to especially thank Emma McKinley for agreeing to write the lead chapter in this volume and for continuing to lead the charge in developing a supportive Marine Social Sciences Community and for Michelle Voyer for reading a draft of, and writing the foreword to this volume.

Maeve Rourke played an instrumental role in assisting the editors and was central to the timely and efficient delivery of the final collated volume to the publishers.

We would like to sincerely thank the Routledge team and their associates for their professionalism, patience, and consistent support, particularly Faye Leering, Prachi Priyanka, and colleagues of the editorial team at Routledge.

This edited volume emerged from Coastal Transitions II: The Blue Economy. This conference was held online in November of 2022 in a virtual environment (due to the COVID-19 pandemic). The rich debate and generous exchanges between participants, including the scientific research community, government officials, NGO representatives, and the business community, were inspiring and transcended any challenges of the virtual format.

The conference, and by extension this edited volume, would not have been possible, or successful, without the key input of Louie Krak, Lauren Brideau, Josephine Lynch, Gabriela Triay, and Michaela Garland. Their tireless efforts are most gratefully acknowledged.

The International Geographical Union (IGU)'s Commission on the Dynamics of Economic Spaces and the IGU Commission on Coastal Systems were key sponsors of Coastal Transitions II: The Blue Economy. Their important support is much appreciated.

This project started as part of the strategic partnership initiative on coastal resilience and transitions in the coastal zone between Liverpool John Moores University, UK, and Southern Connecticut State University, USA. The editors wish to thank those who contributed significantly to developing and nurturing the exchange between our institutions, particularly the international programming staff and university leadership on both sides of the "pond." The editors would also like to acknowledge the role of Mary Immaculate College, University of Limerick Ireland for supporting the project.

It was a great pleasure to work together on this project and we are looking forward to further collaborations in future. C. Patrick Heidkamp would like to personally acknowledge Scott Graves for being a sounding board for ideas and for just being an awesome friend.

Finally, we also would like to thank our families for their endless support and patience!

INTRODUCTION

1

BLUE ECONOMY

People and Regions in Transitions

C. Patrick Heidkamp, John Morrissey, Celine Germond-Duret, and Maeve Rourke

1.1 Coastal Transitions and the Blue Economy

Oceans and coasts are becoming central to the sustainability agenda, as reflected by the SDG14 on "Life below Water." Ocean and coastal economies remain largely untapped frontiers for economic growth, innovation, and development, the arena for new Blue Economies (BEs). This volume comes at an important time for critical perspectives on ocean and coastal development. The United Nations (UN) Decade of Ocean Science for Sustainable Development is projected to run from 2021 to 2030 aiming to bolster the science-policy interface to strengthen collective efforts to manage oceans and coasts worldwide. The Ocean Decade reflects growing interest and investment in the BE, with increasing recognition from bodies such as the European Commission, the UN, and the National Oceanic and Atmospheric Association (NOAA) on the potential for ocean and coastal economies to drive economic growth, innovation, and sustainability. While the BE concept and agenda have been endorsed by a wide range of actors and across a wide range of public and private actors, there has not been a concerted effort at interrogating the multi-scalar complexities of BE development initiatives in academic discourse. There remains an ambiguity around what the BE is, what it encapsulates, and how it is manifested through policies, practices, and in places.

This edited volume is one of the scholarly outputs of *Coastal Transitions II: The Blue Economy*, a conference sponsored by the International Geographical Union's Commission on the Dynamics of Economic Spaces and the Commission on Coastal Systems, the other output being a Themed Section in *The Geographical Journal* (edited by Germond-Duret, C., Morrissey. J., and C. P. Heidkamp 2022 available as early view online at https://rgs-ibg.onlinelibrary.wiley.com/doi/abs/10.1111/geoj.12483).

DOI: 10.4324/9781003280248-2

The conference—the second of a series of conferences focused on sustainability transitions in the coastal zone[1]—was held in November of 2022 in a virtual environment (due to the COVID-19 pandemic). The aim of the conference was to foster direct dialogue and engagement between the scientific research community, government officials, NGO representatives, and the business community through a transdisciplinary discourse focused on exploring the multi-scalar complexities of the BE and its importance for sustainability transitions in the coastal zone.

The conference had over 300 attendees and presenters from academia (scientists, social scientists, arts, humanities), government, and the business community. The conference's focus on the BE—as a subset of factors shaping transitions in the coastal zone—was largely inspired by the important work of Lewis et al. (2019) and LeHeron et al. (2019) who made an important contribution to the understanding of the BE during the first Coastal Transitions Conference in 2017, while literature on the concept was still very nascent.

It is no surprise then that this volume builds upon a previous title in the Dynamics of Economic Spaces Series: *Towards Coastal Sustainability and Resilience* (Heidkamp & Morrissey, 2019) and collates important state-of-the-art perspectives on the BE, applying important geographical and sustainability transition perspectives and highlighting how BE dynamics are situated in regional contexts and shaped by the people who live there; in the process, the aim is to catalyse geographical and transdisciplinary critical scholarship on the theory and practice of the BE.

1.2 Geographical Research in the Context of the Blue Economy

1.2.1 Part 1 Publication Trends: Blue Economy

The past decade has seen a dramatic increase in the discourse and scholarship on the BE (Figure 1.1; Table 1.1). A quick analysis of publication trends using Google Scholar (advanced search) for articles with the exact phrase "Blue Economy" in the title performed on (3-23-2022) shows a substantial and steady increase of BE-related scholarly output over the past 15 years; from 2 mentions in 2007 (2007 was the first time the term appeared in a title referenced by Google Scholar), papers including "Blue Economy" in the title increased to 217 instances in 2021 (Figure 1.1). Using a more restrictive search focused on English language publications and focused on peer-reviewed articles using both *Academic Search Premier* and *Web of Science* (with "Blue Economy" as Keyword, in Title or Keyword, or in Abstract), a corresponding trend in the number of BE-relevant publications is observed.

As geographers, we are of course especially interested in the geography-specific scholarship on the BE and we found (by restricting the *Web of Science* search to geography journals only) that a similar trend could not be witnessed

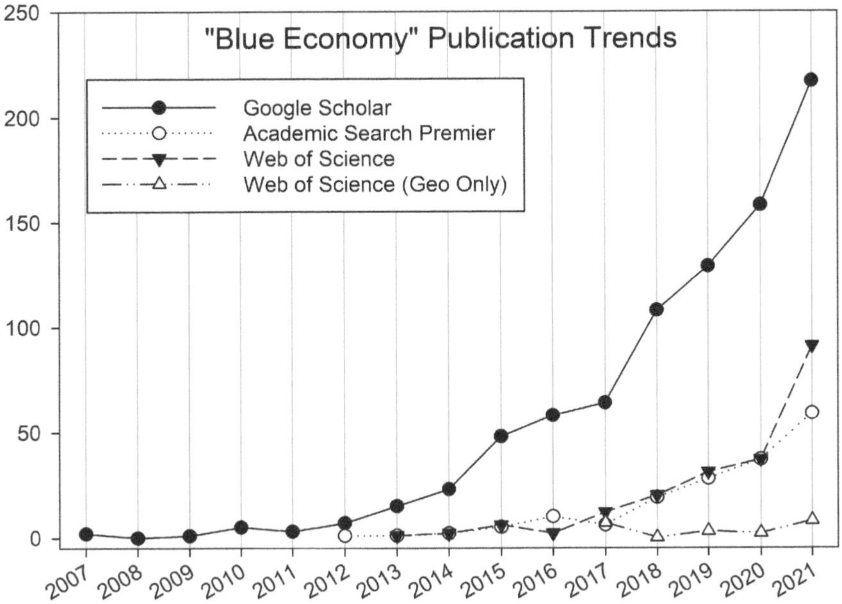

FIGURE 1.1 Blue Economy publication trends 2007–2021

in geography. Aside from a special issue in *Dialogues in Human Geography* (2017), geographers as a community of scholars have had limited engagement with BE concepts, studies, projects, and practices over the corresponding 2007–2021 period.

TABLE 1.1 Blue Economy publication trends 2007–2021

Year	*Google Scholar*	*Academic Search Premier*	*Web of Science*	*Web of Science (geography only)*
2007	2			
2008				
2009	1			
2010	5			
2011	3			
2012	7	1		
2013	15	1	1	
2014	23	2	2	
2015	48	5	6	
2016	58	10	2	
2017	64	6	12	7
2018	108	19	20	0
2019	129	28	31	3
2020	158	37	37	2
2021	217	59	91	8
Total	**838**	**168**	**202**	**20**

As Garland et al. (2019) point out in a meta-analysis of the literature on the BE, there had been limited direct engagement with the BE discourse from an explicitly geographical perspective up until 2019. "Thus far, the existing literature has failed to theorise key geographical concepts such as space, place, scale, and power relations, all of which have the potential to lead to uneven development processes and regional differentiation" (2019, abstract). Notable exceptions are the above-mentioned collection of articles in *Dialogues in Human Geography* published in 2017 (Volume 7, Issue 1) and articles published more recently by some of the authors contributing to this volume (Aschenbrenner & Winder 2019, LeHeron et al. 2019, McKinley et al. 2019).

1.2.2 Importance of the Geographical Context

Part 1 of the volume titled BE People—A Blue Economy for Who? starts with Dr Emma McKinley's chapter which deals with precisely this question and argues for the importance of Marine Social Science perspective(s) on the BE discourse. The other four chapters in Part 1 directly add to this discussion by questioning who in fact benefits from the general BE narrative and from specific BE-focused policies.

Kelly and Axon (Chapter 3) examine the imperative for effective stakeholder engagement approaches in coastal transitions. Their Maximise, Organise, Manage model forwards a novel means of delivering on principles of consensus building, democratic decision-making, and collective action to meaningfully involve stakeholders in coastal transitions processes. Axon (Chapter 4) interrogates the social dynamics around "Blue Energy," focusing on offshore renewable energy projects and the community acceptance of these in varying contexts. The concept of "place-technology fit" is central to the argument of this chapter, applied by Axon to underline the need for tailored local and regional understandings to better inform proposals for renewable energy installation development. Johnson and Hanes (Chapter 5) present a study of conflicts and social acceptance in three aquaculture regions in Maine (USA). This chapter forwards a compelling argument that a nuanced application of marine social science is critical for advancing aquaculture as a key part of the BE. It does so by providing deep insight into the dynamics of conflict and/or acceptance related to local marine aquaculture and into the complex underlying values shaping these dynamics. Kokorsch and Stein (Chapter 6) present a critical analysis of the Finnafjord harbour project in Northeast Iceland, in the process grappling with some of the inherent tensions of BE thinking, including challenges of scale and issues of asymmetrical costs and benefits across environmental, social, and economic domains. This chapter contains valuable insights into resilience-building strategies not only for Iceland but also for coastal communities more widely.

Part 2 of the volume titled BE Regions—Blue Economies in Place showcases a series of chapters aimed at highlighting the importance of place and the regional context in analysing BE policies and initiatives.

Yavan and Kalelioğlu (Chapter 7) apply a robust empirical analysis to assess the potential, performance, and size of the BE in Turkey. This chapter engages with the European Commission's BE framework to determine the extent and spatial distribution pattern of BE activities across the regions of Turkey. Lewis and LeHeron's contribution (Chapter 8) is a continuation of their pioneering work on the BE in Aotearoa New Zealand. This chapter documents the essential work of social scientists in building a knowledge infrastructure for enacting novel environment-economy relations, specifically as part of the New Zealand's Sustainable Seas National Science Challenge. This chapter further underlines the potential of BE thinking to deliver innovative knowledge, social action, and economy. Torres (Chapter 9) examines changing land ownership in Barbuda through a social justice lens. This chapter provides an invaluable perspective on the real-world difficulties of delivering on the promises of BE, including promises of environmental and socioeconomic sustainability, in a regional context of climate-related disruption and economic vulnerability. Winder (Chapter 10) conducts an extensive review of marine spatial planning (MSP) and BE concepts through an entangled governance frame. This chapter provides critical reflections on the potential of both MSP and BE to deliver on Blue Growth and Sustainable Seas aspirations, arguing that these concepts need to be studied together and in place. Winder convincingly posits that attention to entangled planning frameworks and practices can greatly enrich the study of transition and coastal and community resilience challenges. Addressing a key gap in knowledge, Klein, Lisdat, and Tamasy (Chapter 11) provide critical insights on strategic framings, governance mechanisms, and discursive constructions of sustainability within a regional BE setting. Their analysis of the Baltic Sea Region provides a valuable regional perspective, while also offering wider insights into the multi-scalar complexity of developing coherent, situated BE agendas.

Part 3 of the volume, titled BE Futures—Blue Economies in Transition, consists of four chapters which report on research aimed at analysing a shift to a just and sustainable BE from a Geography of Transitions perspective (Coenen et al. 2012, Bridge et al. 2013, Hansen & Coenen 2015). Part 3 of this volumes links explicitly to our previously edited volume published in the same series, *Towards Coastal Resilience and Sustainability* (Heidkamp & Morrissey 2019).

Aschenbrenner and Winder (Chapter 12) investigate governance and decision-making processes in the context of Port futures in Auckland, Aotearoa New Zealand. This chapter analyses the role of ethics in decision-making and questions how this is related to framing and decision-making in BE projects. The comprehensive analysis is based on an extensive stakeholder engagement effort. Providing an important European Union (EU) perspective, the chapter by Aricioglu and Bicer Ucar (Chapter 13) presents a timely review of policies implemented in the EU in order to promote the BE. This chapter focuses specifically on the case study of Blue Growth Clusters in France. Krak and Heidkamp (Chapter 14) tackle the crucial issue of carbon, specifically related to BE as a means to sequester carbon, in their review of "Blue Carbon." The results presented in Chapter 14 show the considerable scope for academic, practical, and policy-related research on this under-studied topic; while most blue carbon research to date centres on mangrove and seagrass carbon capture, references to macroalgae such as seaweeds and kelps are considerably fewer, despite having increased in recent years. Brideau et al. (Chapter 15) address transdisciplinary action research in coastal contexts through a reflective piece on the co-creation of knowledge between industry and academia in an applied water quality, biodiversity, and aquaculture project in Martha's Vineyard, USA. This chapter forwards that transdisciplinary action research can be facilitated and realised through researching novel or applied social and scientific endeavours, supporting shared goals, and via project familiarity among collaborators.

These collective works showcase the BE concept as a potential driver of regionally sensitive, ecologically embedded, and community-focused sustainability. The scope for BE to form a core "cog" in our low-carbon future is obvious, from the potential for renewable energy production and coastal resilience building to possibilities for sustainable food production and the delivery of economic opportunities for peripheral communities. However, fundamental questions remain on how to meaningfully deliver on the sustainability promises of BE, including how to avoid embedding a model of damaging extractivism, as per the terrestrial economy, and how to deliver on the key social sustainability principles of human well-being, equity, and justice when planning and developing BEs. As the UN Decade of Ocean Science for Sustainable Development opens up, this volume provides a timely reminder of the richness, diversity, and potential of coastal and marine spaces. The volume makes an important contribution to the analysis of the BE through furthering geographical and transdisciplinary understandings of BE. It does this by focusing on people (Part 1), places (Part 2), and BE futures (Part 3).

The volume concludes with a call for more critical engagement with the BE by Heidkamp et al. (Chapter 16) and a focus especially towards a consideration of *Diverse Economies*—which will also be the theme of the 3rd Coastal Transitions Conference tentatively scheduled for Fall 2023.

The volume as a whole establishes an important and mutually reinforcing link between the still nascent geographical scholarship on the BE and the larger body of work on the BE in Marine Social Sciences and sets a baseline for continued scholarly engagement with the BE from multiple social science perspectives. We project a significant increase in "BE" research and publications over the next decade and urge for continued consideration of the geographical perspective. As such, this volume is just one step towards more engaged geographical scholarship with the BE.

Note

1 A third conference in the series is currently planned for fall 2023. See http://www.coastaltransitions.org/

Bibliography

Aschenbrenner, M., & Winder, G. M. (2019). Planning for a sustainable marine future? Marine spatial planning in the German exclusive economic zone of the North Sea. Applied Geography, 110, 102050.

Bridge, G., Bouzarovski, S., Bradshaw, M., & Eyre, N. (2013). Geographies of energy transition: Space, place and the low-carbon economy. Energy Policy, 53, 331–340.

Coenen, L., Benneworth, P., & Truffer, B. (2012). Toward a spatial perspective on sustainability transitions. Research Policy, 41(6), 968–979.

Garland, M., Axon, S., Graziano, M., Morrissey, J., & Heidkamp, C. P. (2019). The Blue Economy: Identifying geographic concepts and sensitivities. Geography Compass, 13(7), e12445.

Hansen, T., & Coenen, L. (2015). The geography of sustainability transitions: Review, synthesis and reflections on an emergent research field. Environmental Innovation and Societal Transitions, 17, 92–109.

Heidkamp, C. P., & Morrissey, J. (Eds.). (2019). *Towards coastal resilience and sustainability.* Routledge.

LeHeron, E., Logie, J., Allen, W., Le Heron, R., Blackett, P., Davies, K., … & Hikuroa, D. (2019). Diversity, contestation, participation in Aotearoa New Zealand's multi-use/user marine spaces. Marine Policy, 106, 103536.

Lewis, N. (2019). Cultivating diverse values by rethinking Blue Economy in New Zealand. *In Towards coastal resilience and sustainability* (Heidkamp, C.P. and J. Morrissey, Eds.), Routledge: 94–108.

McKinley, E., Aller-Rojas, O., Hattam, C., Germond-Duret, C., San Martín, I. V., Hopkins, C. R., … & Potts, T. (2019). Charting the course for a Blue Economy in Peru: A research agenda. Environment, Development and Sustainability, 21(5), 2253–2275.

PART 1
Blue Economy—People

2

A BLUE ECONOMY FOR WHOM?

Linking Marine Social Sciences with Blue Economy Discourse

Emma McKinley

2.1 Introduction

Our global ocean, seas and coastlines support some of our most productive and valuable ecosystems, providing resources and benefits for communities, as well as an ever-growing and diverse mix of users and industries. Recent years have seen the ocean, and the concept of a Blue Economy, positioned as a salve, or a solution to various social and economic struggles being faced by nations across the globe. However, achieving a truly sustainable ocean-based economy will be complicated, complex, and multi-faceted. On one hand, the Blue Economy has historically centred on supporting existing and emergent maritime industries, on economic growth, on the creation of jobs and the various elements supporting this. On the other, it is becoming increasingly apparent that this must become a broader conversation taking account of ocean health and societal well-being to deliver and implement a model for sustainable Blue Economy agendas worldwide. This growing recognition has occurred in parallel with the increasing sense of urgency that has come to be associated with the various challenges facing the global ocean—impacts of the climate crisis, ecosystem degradation and biodiversity loss, and the levels of extreme development and industrialisation being experienced by coastal zones across the world. For society to be an active part of Blue Economy dialogue, there is first a need to better understand the complex layers of the relationship between people and the sea and how these can be included as part of the Blue Economy. This is multi-layered, encompassing the multiplicity of values held across different users, stakeholders, and communities and, therefore, must be viewed through multiple lenses—there is therefore a clear, and indeed urgent, role for marine social sciences to contribute to this and for us to consider how human connections to the sea, in all of their dimensions, can become more integrated in global efforts to develop a sustainable and

DOI: 10.4324/9781003280248-4

just Blue Economy. Through this chapter, we will explore how the concept and definitions of the Blue Economy have evolved in recent years, introducing initial thoughts as to how marine social sciences can contribute to this dialogue as we look to the future.

2.2 The Evolution of the Blue Economy

From the outset, it is important to acknowledge that the "Blue Economy," i.e. the role of the ocean in trade and transport, and indeed, everyday society, is not a new or novel construct. Throughout history, the ocean has acted as a connector between people and places, both at the coast and further inland. Although the term and definition remain a point of debate (more on this later), what is evident is that some form of Blue Economy activity or agenda has been present across communities and geographies, space and time, for generations. The activities and industries represented within the "Blue Economy" continue to change and can be mapped across something of a continuum, with those industries now perhaps thought of as being more traditional maritime industries, such as fishing, ports, and shipping, to the emergent industries of biotechnologies, ocean-based renewable energies, mariculture, and more.

While this chapter primarily explores the human dimensions of the Blue Economy, an area which has received relatively little attention until recent years, conversations about achieving or delivering Blue Economy have been, and continue to be, widespread in contemporary ocean governance debate and are commonly found to be high on the policy agenda at a range of scales. We have seen Blue Economy plans or frameworks or agendas set out by nations across the world (the Caribbean, Seychelles, and the European Union (EU), for example) particularly in response to the ever-increasing emphasis of the ocean as a solution to global economic crises.

In 2012, following a period of economic crisis after the financial crash of 2008, the EU set out their roadmap for "blue growth strategy" initiating a conversation which has not quietened since. Much of these discussions, and those from other nations, have centred on the opportunities available to governments and businesses to use the ocean and its resources to stabilise economies, encourage economic growth, provide jobs, and increase employment rates, i.e. the economic benefits of developing a Blue Economy agenda, with little focus on the social and cultural implications, both positive and negative, that this may have. Even in the recently published "A Sustainable Ocean Economy in 2030: Opportunities and Challenges" report, there is relatively little consideration given to the non-economic components of the Blue Economy dialogue (The Economist Group, 2020). However, there is some evidence to suggest the conversation is changing—for example, as a result of the impacts of the COVID-19 global pandemic, 2020 was a year which saw the emergence of the "blue recovery" discourse, with the High Level Panel for a Sustainable Ocean Economy Report (Northrop et al., 2020) stating that while "The ocean economy may be a victim of the impacts of the COVID-19 crisis … it also holds solutions for rebuilding a more resilient, sustainable and equitable post-COVID world.… Investment in 'blue' recovery and stimulus packages, along with policy reform, can immediately create jobs and provide short-term economic relief, all while fostering long-term economic growth, resiliency and social and environmental benefits" (p 1).

There is no sign of this discourse abating, and the facts remain that the ocean ***does*** offer opportunities—however, critics of the Blue Economy concept raise a number of concerns. For example, it is increasingly evident that if we ***only*** view the Blue Economy agenda through an ***economic lens***, we run the risk of degrading our ocean, depleting the ocean resources on which so many of our global communities and industries are dependents, further disconnecting society from the ocean, excluding and marginalising communities, displacing communities, and fundamentally aggravating social inequities and injustices and exacerbating the already numerous "wicked" problems that our ocean faces (Bennett et al. 2019, 2021). Despite this longstanding connection to the ocean, coasts, and seas, at the turn of the 21st century, there was a growing recognition of the disconnect that was becoming increasingly apparent between much of society and the ocean (Jefferson et al., 2014, 2015; McKinley et al., 2020a; Potts et al., 2016). This "sea blindness" or recognition that society, decision makers, and indeed, industry has, in many ways, been looking away from the ocean for a substantial period of time, even when developing policies and management strategies which impact it. This disconnect is now frequently being cited as one of the greatest challenges facing the sustainable future of the ocean across the world (Millennium Ecosystem Assessment, 2005; Pascual et al., 2017). This trend has been mirrored in the evolution of the Blue Economy; indeed, Garland et al. (2019) stress that "perspectives that are economically important, politically central, socially relevant, and environmentally sensitive are currently being omitted in the existing BE discourses." Therefore, as we recognise that the Blue Economy is indeed everywhere, and has been for some time, if we are to deliver a Blue Economy which is equitable, inclusive, and sustainable across a range of indicators, it is both timely and necessary for us to ask, who should be involved in making decisions about the ocean and related Blue Economy initiatives? Who has access to the Blue Economy? Who benefits and who loses? January 2021 saw the launch of the UN Ocean Decade of Ocean Science for Sustainable Development 2021–2030, which explicitly calls for a "transformational relationship between society and the ocean" and clearly states a need for better "integration between natural and social sciences." With the momentum building through the UN Ocean Decade (as well as other international and national obligations and frameworks, including the UN Sustainable Development Goals, for example), we are at a tipping point and the decisions we make now will influence our ocean of the future—therefore, summarising the questions set out earlier in this section, it is crucial that we ask—**Who is the Blue Economy for?**

2.2.1 Defining the Blue Economy

If we are to work towards a more sustainable and inclusive economy for all, we need to be aware of where we're starting from—however, as alluded to in the previous section, defining what we mean by the Blue Economy is not straightforward (Garland et al., 2019; Voyer et al., 2018). Building on earlier discussions on "sustainable development," more recently "green growth," and, latterly, a reframing of this concept for the ocean, the notion of a global Blue Economy emerged from the discourse around the Rio +20 United Nations Conference on Sustainable Development in 2012 (Anon, 2014).

As the discourse has evolved over time, so too have the definitions and frameworks—with many terms becoming interchangeable—"ocean economy," "Blue Economy," and "blue growth" are just a handful of these examples. Table 2.1 presents a selection of some of the more widely recognised definitions set out by key international organisations.

A review of these terms quickly highlights both their commonalities and their differences. In particular, as explored earlier, the focus on economic growth,

TABLE 2.1 Examples of definitions of the blue or ocean economy

United Nations	"The **blue economy** comprises a range of economic sectors and related policies that together determine whether the use of ocean resources is sustainable. An important challenge of the blue economy is to understand and better manage the many aspects of oceanic sustainability, ranging from sustainable fisheries to ecosystem health to preventing pollution. Secondly, the blue economy challenges us to realize that the sustainable management of ocean resources will require collaboration across borders and sectors through a variety of partnerships, and on a scale that has not been previously achieved."
European Commission	"**Blue growth** is the long-term strategy to support sustainable growth in the marine and maritime sectors as a whole. Seas and oceans are drivers for the European economy and have great potential for innovation and growth"
World Bank	"Sustainable use of ocean resources for economic growth, improved livelihoods, and jobs while preserving the health of ocean ecosystem"
The Commonwealth Secretariat	"The '**Blue Economy'** is an emerging concept which encourages better stewardship of our ocean or 'blue' resources. It underpins the thinking behind the Commonwealth Blue Charter, highlighting in particular the close linkages between the ocean, climate change, and the wellbeing of the people of the Commonwealth."—Commonwealth Secretariat (2021)

industrial development, and job creation is notable as the primary driver in all but one of the definitions presented here. Placing the spotlight on economics, and economic valuation of ocean resources, has historically been the norm—further emphasised, perhaps in part, through widespread acceptance of the economic valuation that has tended to go hand in hand with the ecosystem services and, more recently, the natural capital approach dialogue. Of course, it should be noted that although generally accepted by those within ocean governance and management as useful terms, the discourse around these terms and concepts themselves is in a state of flux (see McKinley et al., 2018 for an example of this debate). In contrast to the first three definitions presented in Table 2.1, however, the definition set out by The Commonwealth's Blue Charter (Commonwealth Secretariat, 2018) frames the concept of the Blue Economy as being about more than economic development of ocean-related industries; positioning it as a framework which champions and encourages more integrated and inclusive ways of thinking, linking a number of key areas which are currently drawing policy focus; including the ocean-climate nexus, the link between ocean and human health and well-being, and the concept of well-being (The Commonwealth, 2021). Again, these aspirations of the Blue Economy as a mechanism for facilitating an improved relationship between society and the sea are not particularly novel; an overarching reimagining of the traditional view of ocean economies was recently termed "the new Blue Economy" and guidance on how to achieve mutual benefits of efficiency, sustainability, and stewardship was set out by Spalding (2016). Yet, despite this and the research effort that has been afforded to the Blue Economy, and associated topics, over the last 15 years or so, Box 1 highlights the lack of certainty and standardisation about the term and concept which remains. This complexity was explored in 2018 by Eikeset and colleagues in a Special Issue of Marine Policy, which sought to provide readers with an insight of the myriad definitions and to highlight the challenges and opportunities of each. Some of the obstacles discussed by Eikeset et al. (2018) in their editorial is that without standardisation, without a common understanding of what we mean by the Blue Economy (or whatever your preferred term might be!), there are risks that we may not be working towards the same common goal; there may be issues of misinterpretation and lack of understanding which may prevent effective interdisciplinarity and collaboration, and, finally, there may be variation in understanding where our knowledge and evidence gaps lie. It is important to recognise that the Blue Economy model is an imperfect one; and that the discourse must be encouraged to evolve and improve, bringing in new viewpoints and new voices to the debate. While there may not yet be consensus on the correct terminology, it is undeniable that our ocean supports an increasingly complex network of ocean industries, users, communities, and institutional processes, which encompass a far-reaching network of communities and value chains, as described in the sections above. Moreover, the evolution of the term, and the parallel shifting discourse, is clearly indicating a demand for more integrated thinking in order to realise a truly sustainable and equitable Blue Economy for all (see e.g. Bennett et al., 2019; Campbell et al., 2021).

2.3 Achieving a Balanced Blue Economy—the Role of Marine Social Sciences

Marine social sciences (like its parent field of disciplines, social science) is an overarching umbrella term which brings together a wide range of subjects and approaches including sociology, politics, economics, and human geography among other areas (McKinley et al., 2020b). Fundamentally, marine social sciences encompass any discipline, research, or practice which can help understand and shape diverse values, perceptions, attitudes, and behaviours towards the ocean and have a role to play in ensuring this insight is integrated into decision-making of the ocean, coasts, and seas (Bennett, 2019; Martin et al., 2017; McKinley et al., 2020b). Although it originates from the wider field of social sciences, increasingly the marine social science community is pushing its own boundaries and we are seeing growing interdisciplinarity bringing in creative practice and visual and performing arts-based research and drawing insights from across the humanities disciplines.

In the context of the Blue Economy agenda, a high-level search of the primary literature databases using the terms "Blue Economy," "blue growth," and "ocean economy," will quickly illustrate the growing body of work exploring a wide range of themes, topics, and issues[1]—encompassing all aspects of the Blue Economy, including small-scale fisheries (Cohen et al., 2019), maritime security (Voyer et al., 2018), blue carbon (Steven et al., 2019), marine renewable energy (Li et al., 2020), as well as a rapidly developing field of work exploring the delicate balance between effective ocean governance and the Blue Economy agenda (see; Voyer et al., 2018).

Marine social sciences afford us a diverse suite of lenses through which existing models and frameworks of the Blue Economy can be critiqued and explored (examples summarised in Table 2.2.). An active community of research and practice for many years, the marine social science community has gained a critical mass in recent years; however, its recognition within the wider sphere of ocean science and decision-making is relatively nascent. To really understand the "who" we are talking about within the Blue Economy discussion, we need to look to those areas of research traditionally associated with the social sciences and bring them into Blue Economy spaces and conversations. In a recent Manifesto for Marine Social Sciences, published in 2020, Bavinck and Verrips (2020) stress this, stating that "We [the marine social science community] must critically engage with the Blue Economy process—[we must] study it and engage with protagonists in their respective forums. The rapidly changing political economy of coasts and oceans underscores the need for critical as well as constructive social-science analyses."

TABLE 2.2 Examples of Blue Economy questions that could be (or indeed already have been) answered through marine social science research and disciplines

Field of marine social sciences	*Example of Blue Economy question/topic that can be addressed*
Geography	• Where should future marine renewable energy projects be located, and what are the implications for local communities? • How can the priorities of multiple regions/nations be considered within a transboundary Blue Economy agenda?
Sociology	• How might fishing communities be impacted by changes in the seafood supply and value chain? • How are fishing communities impacted by management interventions, such as marine spatial planning or marine protected area designations?
Economics	• What are the economic benefits of Blue Economy developments (e.g. offshore wind, new port developments etc.) for small-medium businesses within a coastal community?
Psychology	• What is the social acceptability of tidal lagoons? • What are the public attitudes towards decommissioning of oil and gas platforms and associated infrastructure?
Political sciences	• How effective are the governance processes associated within establishing a Blue Economy agenda? • What are the opportunities for meaningful and effective community and stakeholder engagement?
Humanities	• Where has fishing taken place in the past? And how has this influenced the development of coastal communities in a specific region?
Public health	• What are the public health implications of the development of a new port? • What is the health and well-being impact of displacement on fishing communities?
Education	• What are the skills, knowledge, and capacity gaps relating to marine industries in coastal communities? • How do we encourage new entrants and younger people into ocean careers?

Reflecting on the definitions presented in Box 1, however, it is apparent that the Blue Economy discourse has been dominated by economic framing and indicators of success. However, more recently, we have seen researchers explicitly place social science-related topics, and therefore, people and communities, at the heart of their models of a sustainable Blue Economy. For example, the model used by McKinley et al. (2018) presents four quadrats of the Blue Economy, with one focusing specifically on "Social Innovation and Partnership" encompassing the wider issues including, for example, consideration of social values, individual and collective behaviours, inclusion of traditional ecological knowledge, skills, and capacity development, and the need for social innovation to support Blue Economy agendas.

The position of people, and as a result the disciplines encompassed within marine social sciences, has been explored further by Bennett et al. (2019), with particular consideration given to what would be required to deliver a sustainable and just Blue Economy. Their paper sets out a series of key action points, which include (i) prioritisation of sustainability, equity, and social justice as core tenets of any Blue Economy discussions, regardless of scale; (ii) development and implementation of comprehensive, transparent, and effective governance frameworks which are supported by the appropriate legislation and agencies; (iii) appropriate scale guidance for the inclusion of local populations and communities within the Blue Economy agenda; (iv) establishment of processes which ensure equitable distribution of benefits for local populations arising from Blue Economy initiatives, across multiple scales; and, finally, (v) adoption of interdisciplinary approaches to developing Blue Economy agendas and initiatives in order to realise true transformational change, sustainability, and inclusion across multiple geographies and scales of governance. Using these priorities as an initial roadmap, it is clear that marine social sciences have a central role in the operationalisation of these priorities and in addressing historical imbalances that have perhaps been present in earlier Blue Economy discourses.

The UN Ocean Decade, and initiatives aligned with it, appear to offer an opportunity for us to re-evaluate our existing models and approaches to the Blue Economy. The Decade's aspirational calls for a transformed relationship between society and the ocean improved inter-, and indeed, transdisciplinary working across the science-policy-practice interface, and enhanced levels of ocean literacy across all groups of society set out a framework with marine social sciences at its core if it is to succeed. These messages from the UN Ocean Decade represent something of a global shift in the general discourse around how decisions are made about the ocean, who is involved and how they should be implemented. While ocean literacy is not a perfect concept (explored in recent work by McKinley and Burdon, 2020; and in a special issue of the Journal of Marine Frontiers in 2019) it has gained international traction and is being viewed as the mechanism though which transformation can take place. With this in mind, it is therefore pertinent (and timely) for us to explore existing definitions of ocean literacy, drawing insight from other relevant concepts, and consider what these might mean for developing a sustainable, equitable, and inclusive Blue Economy for all. At its simplest, ocean literacy refers to having an "understanding of the ocean's influence on you, and your influence on the ocean." Originating in the early 2000s, the concept has undergone an evolution in the last 5 years or so, with recent research moving away from the knowledge deficit approach of its origins and building on the original seven ocean literacy principles (all of which were very focused on knowing about oceanic processes and the wider science of the ocean). In their 2019 paper, Brennan et al. begin to recognise and explore six dimensions of ocean literacy, including attitudes, behaviour, and communication, while a recent report from McKinley and Burdon (2020) reviews the existing model of ocean literacy, alongside other frameworks of marine citizenship (McKinley and Fletcher, 2012), pro-environmental behaviour change (Stoll-Kleemann, 2019), nature connectedness (Milton, 2002), and more, presenting a series of recommendations of how ocean literacy models could be adapted to become more inclusive of all aspects of societal relationships with the ocean. Drawing insight from this new work on ocean literacy and recognising its platform within the UN Ocean Decade (and other international drivers, including COP 26 and the G7 meetings planned for 2021), there are opportunities for us to consider how ocean literacy and the Blue Economy agendas can be better aligned. So, what do we mean when we talk about the Blue Economy agenda through the lens of ocean literacy and wider marine social sciences? If we take the ocean literacy dimension of Knowledge, for example, this could be about ensuring younger members of the community are aware of marine career prospects and have access to the knowledge, skills, and capacity development opportunities to facilitate their engagement with the Blue Economy, in this way. If we explore the dimension of Activism, as defined by Brennan et al. (2019), from a Blue Economy perspective, this could mean an individual voting for policies that are in support of a sustainable and just Blue Economy, e.g. creating new jobs in the maritime sector; while if we explore the dimension of awareness and attitudes, this can relate to the diverse perceptions, social values, and connection that communities have with the ocean and how these might be impacted by a specific Blue Economy initiative. While this chapter in no way presents an exhaustive examination of the links between the Blue Economy and the ocean literacy, or wider marine social science agenda, the opportunities to better align these two agendas, and indeed the mutual benefits that would be realised, are clear.

Of course, there are challenges. If we are to realise the benefits of the insight, skills, and expertise that the marine social science community can bring to Blue Economy discourses, we have to first work collaboratively to address the misunderstandings and lack of awareness of the value and impact of marine social sciences—that is, we need to demystify marine social sciences, shout about success stories, and lessons learned, ensuring it is accessible to all as a core component of an interdisciplinary Blue Economy agenda. One aspect will be increasing capacity for marine social science research and its application, in all its guises, among researchers, practitioners, and decision makers. We need to look to the existing work on the Blue Economy and consider where our evidence gaps are—what do we already know? What has worked? What do we need to prioritise in order to ensure a transition to a just, inclusive, equitable, and sustainable Blue Economy?

2.4 Summary and Concluding Remarks

This chapter has explored the evolutionary journey of the concept of a Blue Economy—and while it is evident that the concept is not new or unique, what is perhaps more novel (and promising) is the growing emphasis on the need for the Blue Economy agenda to take account of the people that these agendas and initiatives impact. In doing so, in connecting society more effectively to the ocean, coasts, and seas, we have an opportunity to realise a more sustainable and inclusive Blue Economy, across economic, ecological, social, and cultural scales. To achieve this, this chapter sets out the following concluding comments:

- There is a need to broaden the "toolkit" and approaches that are being used within the Blue Economy Agenda, drawing on insight, expertise and skills from across the widest possible spectrum of marine sciences. Crucially, Marine social sciences need to be meaningfully integrated into the wider Blue Economy dialogue. Marine social science researchers can, and indeed in many cases already do, provide valuable insight into a vast range of topics which are directly related to the Blue Economy including (but not limited to):
    - Investigation into current issues relating to health, well-being, safety, and security of maritime/coastal workers and communities.
    - Skills gaps, capacity building, and understanding the role of social capital in delivering a sustainable Blue Economy for the future.
    - Understanding and consideration of diverse values and knowledge to support a sustainable Blue Economy.
    - Evaluation of the relationship between the sea, coast, islands, and individual or community sense of identity and place, and how this might be influenced by a Blue Economy agenda across diverse geographies.
    - Exploration of the link between sustainable Blue Economy initiatives and the health and well-being of the ocean ***and*** communities.
    - Interrogation of existing governance and management processes and interventions, understanding their impacts and implications for communities, and identifying how these could be improved to ensure a more sustainable, inclusive, and equitable global Blue Economy agenda.
    - Exploration of trade-offs within decision-making which may result due to Blue Economy initiatives.
    - Promotion and advocation for equal access, rights, and benefits that are derived from Blue Economy developments (e.g. changes in industry within a local community, development of an offshore wind farm, gender equity, meaningful inclusion of coastal communities, and indigenous people within decision-making) to ensure inclusivity and equity.
- We need to learn from what has gone before, evaluate what is (and isn't!) working in the present, and look to the future. The landscape of the Blue Economy is not static—it is dynamic, and ever-changing—we therefore must be flexible and adaptive so we can respond to changing circumstances, evolving industries and temporal and spatial variations in priorities, needs, and decision-making. Furthermore, as communities across the globe continue to recover and respond to various challenges, including ongoing threats of climate change, and at the time of writing, the COVID-19 pandemic, there is an opportunity for us to reframe what we mean by "blue recovery." The rhetoric should not be limited to economic recovery but must be broadened to include ocean recovery (i.e. recovery and restoration

of our ocean and coastal environments to support improved ocean health), as well as recovery of our relationship with the ocean. This recovery can be extended to include social and cultural recovery—and a well thought out, inclusive, socially just model for a sustainable Blue Economy has the potential to deliver these mutual benefits for both the ocean and society.

- Interdisciplinary collaborations across science-policy-practice, bringing in the wider ocean science research community, industry, and decision makers will be integral to achieving sustainable use of the ocean now and in the future. There is no-one-size-fits-all approach to the Blue Economy agenda—and marine social science research and practice are CORE to achieving a just, sustainable, inclusive, and equitable Blue Economy which recognises the specific needs, priorities, and challenges of environments, people, and places at a range of scales. ***However***, currently, there is limited evidence of any ocean governance processes or approaches, at any scale, which easily facilitate the inclusion of marine social sciences evidence and data. While ocean decision-making has historically been dominated by natural and physical sciences, and perhaps the field of economics, there is evidence that the governance landscape is changing—see for example through the link between Wales' Wellbeing of Future Generations Act (2015) and the development of the Wales National Marine Plan, or the establishment of a marine social science team within the UK's Department of Environment, Food and Rural Affairs (Defra) in 2018. Despite this progress, there remains a need for effective pathways to impact for social science data and evidence to be included within decision-making and governance to support development and delivery of a sustainable Blue Economy agenda across multiple scales.
- Finally, there is an opportunity to look to existing networks and communities of research and practice (e.g. the Marine Social Science Network, the MARE consortium, the UK's CPN, and more) to support development of best practice, effective collaboration, and meaningful engagement with communities and stakeholders on issues relating to the Blue Economy.

Note

1 When limited to articles published since 2017, over 17,000 results were found in Google Scholar.

References

Anon. 2014. Blue Economy Concept Paper. United Nations Sustainable Development Knowledge Platform, Department of Social and Economic Affairs. https://sustainabledevelopment.un.org/index.php?page=view&type=111&nr=2978&menu=35

Bavinck, M., and Verrips, J. 2020. Manifesto for the marine social sciences. Maritime Studies 19: 121–123. https://doi.org/10.1007/s40152-020-00179-x

Bennett, N.J. 2019. Marine social science for the peopled seas. Coastal Management 47(2): 244–252. https://doi.org/10.1080/08920753.2019.1564958

Bennett, N.J. Cisneros-Montemayor, A.M., Blythe, J., Silver, J.J., Singh, G. Andrews, N., Calò, A., Christie, P., Di Franco, A., Finkbeiner, E.M., Gelcich, S., Guidetti, P., Harper, S., Hotte, N., Kittinger, J.N., Le Billon, P., Lister, J., Lopez de la Lama, R., McKinley, E., Scholtens, J., Solås, A.-M., Sowman, M., Talloni-Álvarez, N., Teh, L.C.L, Voyer, M., and Sumaila, U.R. 2019. Towards a sustainable and equitable Blue Economy. Nature Sustainability 1–3. https://doi.org/10.1038/s41893-019-0404-1

Bennett, N.J., Blythe, J., White, C.S., and Campero, C. 2021. Blue growth and blue justice: Ten risks and solutions for the ocean economy. Marine Policy 125. https://doi.org/10.1016/j.marpol.2020.104387

Brennan, C., Ashley, M., and Molloy, O. 2019. A system dynamics approach to increasing ocean literacy. Frontiers in Marine Science 6: 360. https://doi.org/10.3389/fmars.2019.00360

Campbell, L.M., Fairbanks, L., Murray, G., Stoll, J.S., D'Anna, L., and Bingham, J. 2021. From Blue Economy to blue communities: Reorienting aquaculture expansion for community wellbeing. Marine Policy 124. https://doi.org/10.1016/j.marpol.2020.104361

Cohen, P.J., Allison, E.H., Andrew, N.L., Cinner, J., Evans, L.S., Fabinyi, M., Garces, L.R., Hall, S.J., Hicks, C.C., Hughes, T.P., Jentoft, S., Mills, D.J., Masu, R., Mbaru, E.K., and Ratner, B.D. 2019. Securing a just space for small-scale fisheries in the Blue Economy. Frontiers in Marine Science 6:171. https://doi.org/10.3389/fmars.2019.00171

Commonwealth Secretariat. 2018. Commonwealth Blue Charter Shared Values, Shared Ocean a Commonwealth Commitment to Work Together to Protect and Manage Our Ocean. Available from: https://bluecharter.thecommonwealth.org/

Eikeset, A., Mazzarella, A.B., Davíðsdóttir, B., Klinger, D.H., Levin, S.A., Rovenskaya, E., and Stenseth, N.C. 2018. What is blue growth? The semantics of "Sustainable Development" of marine environments. Marine Policy 87: 177–179. https://doi.org/10.1016/j.marpol.2017.10.019

European Commission. 2012. Blue Growth: Opportunities for Marine and Maritime Sustainable Growth. Communication from the Commission to the European Parliament, The Council, The European Economic and Social Committee and the Committee of the Regions.

Jefferson, R., McKinley, E., Capstick, S., Fletcher, S., Griffin, H., and Milanese, M. 2015. Understanding audiences: Making public perceptions research matter to marine conservation. Ocean & Coastal Management 115: 61–70.

Jefferson, R.L., Bailey, I., Laffoley, D.d'A., Richards, J.P., and Attrill, M.J. 2014. Public perceptions of the UK marine environment. Marine Policy 43: 327–337.

Garland, M., Axon, S., Graziano, M., Morrissey, J., Heidkamp, C.P. 2019. The Blue Economy: Identifying geographic concepts and sensitivities. Geography Compass 13: e12445. https://doi.org/10.1111/gec3.12445

Li, L., Ruzzo, C., Collu, M., Gao, Y., Failla, G., and Arena, F. 2020. Analysis of the coupled dynamic response of an offshore floating multi-purpose platform for the Blue Economy. Ocean Engineering 217: 107943. https://doi.org/10.1016/j.oceaneng.2020.107943

Martin, V.Y., Weiler, B., Reis, A., Dimmock, K., and Scherrer, P. 2017. 'Doing the right thing': How social science can help foster pro-environmental behaviour change in marine protected areas. Marine Policy 81: 236–246. https://doi.org/10.1016/j.marpol.2017.04.001

McKinley, E., Aller-Rojas, O., Hattam, C., Germond-Duret, C., San Martín, I.V., Hopkins, C.R., Aponte, H., and Potts, T., 2019. Charting the course for a blue economy in Peru: a research agenda. Environment, Development and Sustainability, 21(5): 2253–2275.

McKinley, E., Acott, T.G., and Yates, K.L. 2020b. Marine social sciences: Looking towards a sustainable future. Environmental Science and Policy 108: 85–92. https://doi.org/10.1016/j.envsci.2020.03.015

McKinley, E., Aller-Rojas, O., Hattam, C., Germond-Duret, C., Vicuña San Martín. I., Hopkins, C.R., Aponte, H., and Potts, T. 2018. Charting the course for a Blue Economy in Peru: A research agenda. Environment, Development and Sustainability 1–23.

McKinley, E. and Burdon, D., 2020. Understanding Ocean Literacy and Ocean Climate-Related Behaviour Change in the UK-Work Package 1: evidence Synthesis. Hull: Daryl Burdon Ltd. Available online at: https://darylburdon. co. uk

McKinley, E. and Fletcher, S., 2012. Improving marine environmental health through marine citizenship: a call for debate. Marine Policy, 36(3): 839–843.

Millennium Ecosystem Assessment. 2005. Ecosystems and Human Well-being: Synthesis. Island Press, Washington, DC.

Milton, K. 2002. Loving Nature: Towards an Ecology of Emotion. Routledge, London.

Northrop, E., et al. 2020. A Sustainable and Equitable Blue Recovery to the COVID-19 Crisis. Report. World Resources Institute, Washington, DC. Available online at http://www.oceanpanel.org/bluerecovery

Pascual, U., Balvanera, P., Díaz, S., Pataki, G., Roth, E., Stenseke, M., Watson, R.T., Dessane, E.B., Islar, M., Kelemen, E., Maris, V., Quaas, M., Subramanian, S.M., Wittmer, H., Adlan, A., Ahn, S.E., Al-Hafedh, Y.S., Amankwah, E., Asah, S.T., Berry, P., Bilgin, A., Breslow, S.J., Bullock, C., Cáceres, D., Daly-Hassen, H., Figueroa, E., Golden, C.D., Gómez-Baggethun, E., González-Jiménez, D., Houdet, J., Keune, H., Kumar, R., Ma, K., May, P.H., Mead, A., O'Farrell, P., Pandit, R., Pengue, W., Pichis-Madruga, R., Popa, F., Preston, S., Pacheco-Balanza, D., Saarikoski, H., Strassburg, B.B., van den Belt, M., Verma, M., Wickson, F., Yagi, N. 2017. Valuing nature's contributions to people: the IPBES approach. Current Opinion in Environmental Sustainability 26–27: 7–16.

Potts, T., Pita, C., O'Higgins, T., and Mee, L. 2016. Who cares? European attitudes towards marine and coastal environments. Marine Policy 72: 59–66.

Spalding, M.J. 2016. The new Blue Economy: The future of sustainability. Journal of Ocean and Coastal Economics 2(2), Article 8. https://doi.org/10.15351/2373-8456.1052

Steven, A.D.L., Vanderklift, M.A., and Bohler-Muller, N. 2019. A new narrative for the Blue Economy and Blue Carbon. Journal of the Indian Ocean Region 15(2): 123–128. https://doi.org/10.1080/19480881.2019.1625215

Stoll-Kleemann, S. 2019. Feasible options for behavior change toward more effective ocean literacy: A systematic review. Frontiers in Marine Science 6: 273.

The Commonwealth. 2021. The Blue Economy. https://thecommonwealth.org/blue-economy

The Economist Group World Ocean Initiative. 2020. A sustainable ocean economy in 2030: Opportunities and Challenges. Available here: https://ocean.economist.com/blue-finance/articles/a-sustainable-ocean-economy-in-2030-opportunities-and-challenges

Voyer, M., Quirk, G., McIlgorm, A., and Azmi, K. 2018. Shades of blue: What do competing interpretations of the Blue Economy mean for oceans governance? Journal of Environmental Policy & Planning, 20(5): 595–616. https://doi.org/10.1080/1523908X.2018.1473153

3

EFFECTIVE STAKEHOLDER ENGAGEMENT IN COASTAL TRANSITIONS

Floating away from the DAD Approach towards the MOM Method

Miriah Kelly and Stephen Axon

3.1 Introduction: Stakeholders, Transitions, and the Coastal Zone

Given the economic and environmental value of coastal landscapes, the regional climate impacts on coasts will be significant (Jurjonas and Seekamp, 2018; IPCC, 2021). Expected impacts include sea level rise, storm surge, changes in precipitation, hypoxic and anoxic ocean waters, warmer water temperatures, ocean acidification, and changing ocean currents. These expected impacts are anticipated to become more severe over the next few decades with the new IPCC AR6 indicating that climate change will exacerbate the problems that many coastal communities, towns, and cities already face including shoreline erosion, coastal flooding, and water pollution (Jurjonas and Seekamp, 2018; IPCC, 2021).

In the opening chapter McKinley (2022) asserts clearly that we must seek to understand relationships between stakeholders and the sea, foster interdisciplinary collaboration among diverse stakeholders, and attend to the long-term nature of blue economic growth, and the role of human society in achieving sustainability goals. Yet, previous studies have demonstrated that there is minimal meaningful engagement amongst key stakeholders and communities where large-scale infrastructure and economic development projects are established in the coastal zone (Berkes, 2015; Kelly, 2018). Recent examples of failed offshore wind energy projects include Cape Wind, MA (Boston Globe, 2015), Lake Michigan Wind Farm, MI (Grand Rapids Business Journal, 2010), and Ocean City, MD (The Baltimore Sun, 2018). This leads to substantial opposition to such projects, irrespective of their environmental or economic objectives, which has resulted in failed schemes such as Cape Wind, Massachusetts (Boston Globe, 2015).

It is clear that top-down decision-making practices attempt to circumvent opinions of acceptability and opposition to coastal transition developments. With

DOI: 10.4324/9781003280248-5

limited understanding and acceptability, this results in stakeholders unable to be invested, or participate, in sustainability-related projects (Axon, 2019). What is missing is clear: effective stakeholder engagement. Integrating the principles and practices of meaningful involvement with development schemes in the coastal zone could lead to increased support while also enhancing sustainability and resilience in the coastal zone.

Given the need for multiple responses to address multiple and interrelated environmental and socio-economic challenges at the coastal zone, it is imperative that effective stakeholder engagement approaches are implemented to ensure the principles of consensus, democratic decision-making, and collective action are implemented. Current stakeholder and public engagement approaches largely utilise consultative methods and other forms of tokenistic involvement (Morrison and Dearden, 2013; Axon, 2019). Specific forms of tokenistic forms of stakeholder engagement correspond to decide, announce, and defend (DAD) models, hereafter referred to as the DAD model (Susskind and Elliot, 1983; Dietz and Stern, 2008). We critique this model of stakeholder engagement and highlight its shortcomings to effectively enhance participation in coastal transitions. Following this, we present an alternative approach to this model; identified as the maximise, organise, and manage method, hereafter referred to as the MOM Method. Identifying and describing this alternative approach, we present the theoretical foundations of the MOM method and outline its practical applications in a coastal transition context.

This chapter therefore attempts to draw attention to the value of effective stakeholder engagement principles and practices in the coastal zone. In so doing, we advocate for changes to policy and practice to enable the meaningful involvement of stakeholders in coastal transitions. We, therefore, identify this work to be part of a starting point for future academic, policy, and practitioner-related discussions, reflections, and evaluations of highly involved approaches of stakeholder engagement, including the MOM method as outlined in this chapter.

3.2 Effective Stakeholder Engagement: Principles and Practices

The literature on public participation and environmental decision-making is rich (Depoe et al., 2004; Dietz and Stern, 2008; Cox, 2010). Cox (2010) defines public participation as the ability of individual citizens and groups to influence environmental decisions through (1) access to relevant information, (2) public comments to the agency responsible for a decision, and 3) the right, through the courts, to hold public agencies and businesses accountable for their environmental decisions and behaviours. Depoe et al. (2004: 115) define public participation as "pre-decisional communication between an agency or organization responsible for a decision and that organization's relevant public community." Both of these definitions presented by key authors in this area of research focus on the inclusion of appropriate stakeholders in the process of making decisions that have the potential to affect them. They also assert that citizens have the right, ability, and legal authority to engage in a process that has the potential to shape their futures or impact their lives in some way.

To this end, Depoe et al. (2004) emphasise the importance of meaningful public participation, noting that the need for public participation extends well beyond what is required by law (e.g. in the National Environmental Policy Act). The one way, information dissemination model of public participation, often used by government agencies is insufficient, as it lacks attention to the valuable ideas and opinions the public holds. Instead, a two-way model of public participation that fosters dialogue and a meaningful exchange of information is required to truly honour the value stakeholder participants have in natural resource decision-making (Depoe et al., 2004; Cox, 2010).

When done properly, public engagement can result in a more comprehensive understanding of an issue, broaden the options that can be considered for addressing the issue, and improve the acceptability of policy decisions (Dietz and Stern, 2008). However, it often falls short because it occurs too late in the process, fosters a more adversarial approach, doesn't provide the space for meaningful dialogue, and lacks mechanisms for ensuring public values are accounted for (Depoe et al., 2004). As conventional public participation processes have revealed limitations, agencies, communities, and stakeholders have turned to more collaborative approaches (Kelly, 2018).

3.3 The DAD Approach: Decide, Announce, and Defend

The majority of stakeholder engagement activities are typically approached by applying consultative exercises (Morrison and Dearden, 2013). Consequently, this form of involvement in policy-making or deliberative decision-making attempts to reassure stakeholders and/or communities that their "voice" is being heard. Arnstein (1969) asserts that citizen participation, and stakeholder engagement by extension, is a categorical term for citizen power, more specifically; it is the redistribution of power that enables citizens and stakeholders to be deliberately included in political and economic processes. From this foundation, Arnstein (1969) presents a "ladder of citizen participation" that reflects a typology of participation and non-participation alike that corresponds to the extent of stakeholders' power in determining end results (Figure 3.1).

The bottom rungs of the ladder reflect "manipulation" and "therapy" that correspond to non-participation in deliberative decision-making processes. The real objectives of these bottom rungs are not to enable genuine participation but enable power-holders to "educate" or "cure" stakeholders (Arnstein, 1969). However, tokenistic forms of participation allow stakeholders to have a voice, though any contributions are unlikely to influence the final result. This gives rise to the decide, announce, and defend (DAD) model of stakeholder engagement, referred to as decide, announce, and defend exercises for participation. This approach has been well-applied in coastal sustainability transitions and relies

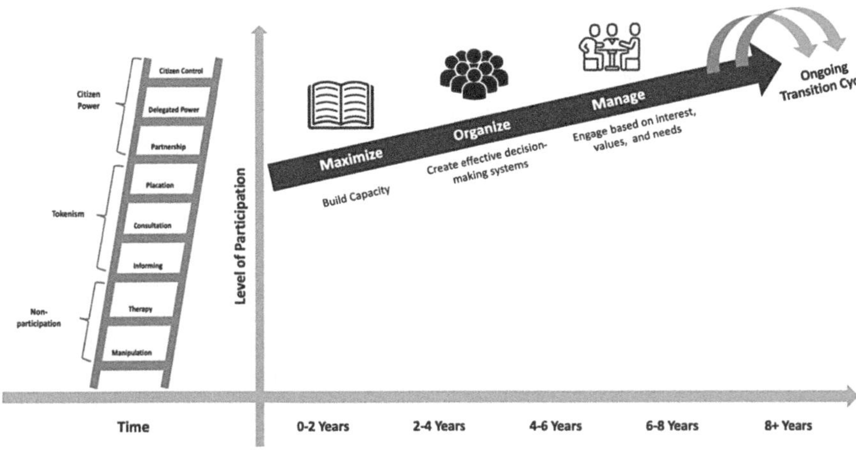

FIGURE 3.1 A ladder of citizen participation

heavily on a top-down method that limits meaningful opportunities for stakeholder engagement. The DAD model therefore relies on expeditiously soliciting opinions and tokenistically responding to concerns of proposals or developments, thus separating itself from the higher rungs of Arnstein's (1969) ladder. These higher rungs represent citizen power where increasing degrees of decision-making clout with key stakeholders obtain control of the majority of the decision-making process. While this ladder represents an over-simplification of stakeholder engagement in coastal transitions, we find this to be a useful framework for contextualising the DAD model and its tokenistic objectives predicted on consultation and placation approaches.

Despite this, the DAD model and its largely consultative exercises can be useful in particular circumstances. From a purely time management perspective, it is often the case that legislative and development timelines require engaging stakeholders quickly to reduce exorbitant financial pressures that challenge transition processes. This is of particular importance where local, regional, or national governmental changes occur that may impact future trajectories of the projects. Consequently, the value of the DAD model of stakeholder engagement is that it can be applied quickly to ensure awareness of transition processes, outcomes, and likely disruptions in a limited time window.

Despite its ubiquity and ability to be performed in short time frames, the DAD model reflects a short-term outlook to stakeholder engagement that is applied expeditiously and rarely accounts for understanding the subtleties of individual, and community, specific perspectives in great detail. Rather, consultative practices often rely on a purely quantitative measure of support versus opposition to measure the potential success of coastal transition processes and developments. Satisfying the immediate needs of local governments and institutions' strategic objectives through applying the DAD model comes at the cost of ignoring the underlying factors that shape support and opposition, thus relying on an over-simplified understanding of stakeholder perspectives that neglect the specificities of coastal cultures, communities, and places. As a result, we argue here that this short-term approach clearly reflects engrained institutional practices that demand prompt involvement rather than meaningful engagement and longer term, strategic vision that continuously engages stakeholders effectively.

3.4 The MOM Method: Maximise, Organise, and Manage

The maximise, organise, and manage (MOM) method that we outline in this chapter is an approach that can be employed for improved stakeholder engagement in coastal communities to ensure just transitions are effectively implemented. Though MOM-like efforts take more time, and require more substantive effort by all parties involved, the result is a deeper and more strategic approach that allows the "voice" of stakeholders and communities to be heard (Axon, 2019). Furthermore, this is an effective way to meaningfully engage stakeholders through longer time periods in broader coastal spaces. Boundary organisations have a key role in serving as long-term support for meaningful stakeholder engagement over longer periods of time. Having an established entity to support long-term stakeholder engagement is essential to implementing the MOM method, and maintaining strong relationships with stakeholders and their social networks is an important consideration.

Tribbia and Moser (2008) suggest that simply providing "more information" and "better information" are insufficient and that there is a distinct need for boundary organisations to broker information between scientists and end users. Boundary organisations can help improve the end-to-end process of knowledge co-production and application by enabling scientists and decision-makers to increase mutual understanding of capacities and needs while remaining within their respective professional boundaries (Cash et al., 2003). Moving beyond simply providing information to residents and other stakeholders in coastal communities where transitions may, and will, occur requires meaningful involvement. We outline in the proceeding sections just how this can be achieved.

3.4.1 Maximising Stakeholder Capacity to Participate

Coastal spaces are characterised by the complex incremental changes that occur over longer periods of time. The MOM method is about developing an approach that acknowledges ways stakeholders can be meaningfully engaged throughout longer periods of time. It begins with "maximizing stakeholder capacity." In this first phase, focusing on building capacity and empowering stakeholders to participate in environmental decision-making is essential. The oft-employed tokenistic information deficit-based approach is not, however, adequate as a technique of building capacity (Tribbia and Moser, 2008; Axon, 2019;

Jurjonas et al., 2020). Rather, the MOM method encourages the use of robust stakeholder assessment to understand the interests and values of possible participants (Jurjonas et al., 2020). Given that coastscapes have deeply embedded social meanings and shape local economies and identities (McLachlan, 2009; Poe et al., 2014), clear understanding of the ways in which coastal and marine areas are used is essential to prevent opposition to proposed transition development.

These assessments can then be situated within the broader context of transition development to create more meaningful capacity-building efforts that take into account the values and interests of the participants, and not just what decision-makers assume is required from purely econometric perspectives. For example, Resilient Connecticut, a two-county project (in the State of Connecticut, USA) designed to build capacity of coastal communities to adapt to sea level rise, has dedicated significant time and effort to better understand and assess stakeholders. In this program, Resilient Connecticut have sought to uncover not just perceived information needs of different stakeholder groups, but also how they prioritise climate topics, and ways in which they prefer to be engaged or involved. It is important to note here that diversifying the stakeholder pool must begin at this stage to ensure that transitions do not cater for specific sub-groups of coastal communities and must instead attend to the values, interests, and needs of all groups represented at the coast (Carley and Konisky, 2020) as well as identifying specific vulnerabilities and adaptive capacities (Jurjonas et al., 2020). If underserved and underrepresented communities are not involved in capacity building in an equity-based way, then just transitions in the coastal zone are unlikely to occur (Few et al., 2011; Carley and Konisky, 2020; Jurjonas et al., 2020).

3.4.2 Organising Stakeholders into Existing Systems and Sub-systems

The next phase of the MOM method, involves organising stakeholders into existing systems and subsystems. Often public and stakeholder engagement efforts seek to establish new committees, boards, and organisational entities to manifest the desired work. This approach does not take into account the longevity that existing organisational systems already have (i.e. wetlands commissions, organisational leadership boards, etc.). Rather than attempting to develop new systems of organisation, those interested in engaging stakeholders with transition development should integrate stakeholder groups who have built capacity in phase one into existing systems.

When multiple individuals from the same organisational system (e.g. numerous members from the same municipality—water sewer managers, land-use planning, health districts, etc.) collaborate and incorporate the values, interests, and adaptive capacities of coastal communities, transitions are able to develop effectively towards shared visions and objectives (Kelly et al., 2018; Carley and Konisky, 2020; Jurjonas et al., 2020).

Building connections within and between systems are both important for any systems-based approach, particularly given that coastal areas are used for a variety of recreational, cultural, industrial, and educational purposes by diverse populations, actors, and industries (Poe et al., 2014). Yet, the focus of many stakeholder engagement efforts is around connecting institutional entities, rather than building the strength of existing organisations. Applying a systems-based approach tool like stakeholder systems mapping can be effective for understanding the existing organisational identities different stakeholders maintain and how this can fit in, and extend, existing systems and sub-systems to enhance coastal transitions.

3.4.3 Managing Stakeholders through Transition Cycles

Knowing when, and when to not, involve stakeholders is key to effective management of stakeholders in a just transition context. Local adaptation relies heavily on decision-making. Few et al. (2011) emphasise the importance of involving stakeholders in meaningful ways. It is important to realise that constant participation is not necessarily the most effective communication strategy for managing stakeholder engagement over time given that people can move from active involvement to burnout rapidly. Rather, being thoughtful and considerate in requesting meaningful stakeholder participation in distinct and strategic ways is more effective (Axon, 2019). Transitions cycles, by their nature, are just that; "cycles" and all stakeholders don't necessarily need or want to be involved in all phases of the cycle.

Additionally, various methods of engagement could include, but are not limited to, holding membership of decision-making panels; holding adaptive capacity assessment discussions with policy and decision makers; engaging local coastal communities with transition processes; meaningfully involving local tribes and Indigenous communities who are frequently overlooked with transition cycles; and being available to respond to action plans and attend meetings to ensure local community voices are heard. Axon (2019) outlines that one of the main lessons for public and stakeholder engagement is that involvement and participation in coastal transition processes is subject to temporal variation and is not fixed. Rather, involvement needs to be facilitated in the short term and sustained in the long term should "active" engagement in coastal transitions be maintained (Axon, 2019).

As shown in Figure 3.2, the MOM approach is a long-term model for integrating stakeholder interests in meaningful ways. Note that the maximising stakeholder capacity stage bridges the gap between tokenistic and powerful stakeholder engagement. Using enthusiasm or interest as a foundation for co-developing knowledge and skills is key to maximising partner interests during the early years of engagement. Further, once co-developed knowledge is constructed, developing decision-making processes and structures that relate to the specific context of the situation is key to maintaining an adaptive organisational approach. Doing so is crucial for the coastal transition context which, as McKinley explains in

TABLE 3.1 MOM approach phases and applications to coastal transitions, marine social science, and the Blue Economy

Phase	*Best practices*	*Coastal transitions applications*	*Value to marine social science*	*Blue Economy example*
Maximising stakeholder capacity to participate	Seek to understand stakeholder interests Collect data that are meaningful to a variety of groups Be willing to openly share information	As coastal communities change, along with values and beliefs associated with coastal environments, being thoughtful and purposeful in understanding the diversity of coastal stakeholders is key	McKinley (2022) calls for an evaluation of the relationships people have with the sea. Understanding the diversity of ocean identity perspectives in diverse coastal communities is essential to effective engagement	In the context of offshore wind development, deeply understanding stakeholder perspectives can aid in decision-making around offshore wind infrastructure citing, management, and it can minimise environmental impact implications
Organising stakeholders into existing systems and sub-systems	Think creatively about group structure based on interests and skills Provide opportunities for stakeholders to contribute to the process or the outcome Focus on relationship building as a goal	We posit that integrating organisational systems is a mechanism for developing modern and innovative decision-making processes specific to coastal geographies and contexts, and we note that terrestrial models of decision-making may not be appropriate for the coastal context	McKinley (2022) explains interdisciplinary collaboration is key to structuring sustainable coastal transitions. Marine social science holds promise as a field for enacting a transdisciplinary community of academics and practitioners	In the context of coastal port development, having decision-making structures in place that integrate broader stakeholder values, and are reflective of a greater regional area, is pivotal to sustainable port management, and contributes to the Blue Economy
Managing stakeholders through transition cycles	Create simple stakeholder information management systems Use technology to manage stakeholders Continue to foster relationship building over time	Maintaining stakeholder interest and engagement over years, or even decades is difficult. Yet, having systems in place to engage stakeholder capacity at different points in time based on needs and interest is vital to mirroring the long-term trajectory of stakeholder engagement across blue growth contexts	McKinley (2022) explains blue growth is a long-term strategy/goal. Hence, our approaches to stakeholder engagement must rely on our understating of social and physical environmental changes over time	In the context of aquaculture, enacting this model would allow for strategic support for sustainable aquaculture growth—in citing aquaculture operations, monitoring conditions, and pivoting management when needed, and creating innovative solutions to complex problems

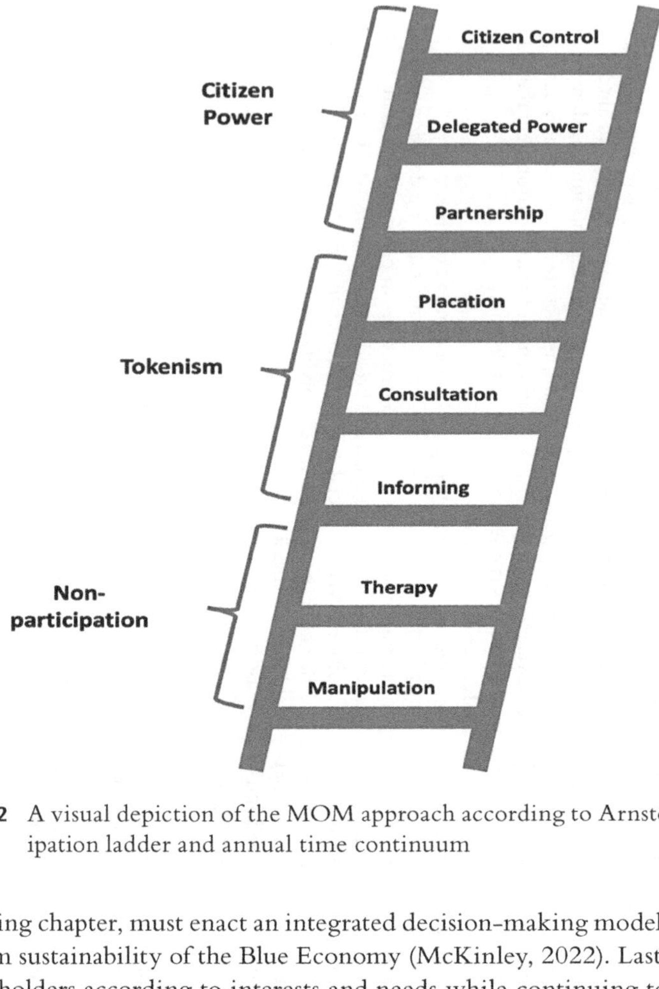

FIGURE 3.2 A visual depiction of the MOM approach according to Arnstein's participation ladder and annual time continuum

the opening chapter, must enact an integrated decision-making model to achieve long-term sustainability of the Blue Economy (McKinley, 2022). Lastly, managing stakeholders according to interests and needs while continuing to empower leadership in making decisions and participating in the process is a mechanism for maintaining long-term participation. Rather than relying on consistent regular engagement over a period of years (i.e. monthly meetings), which may or may not have a distinct purpose or goal, meaningful engagement should rely on a strategic approach to engaging particular stakeholders based on different interests and needs over time, with an emphasis on relationship building and continued collaboration for the purpose of co-developing knowledge that will inform important decision-making related to the coastal zone.

3.5 Conclusions and Recommendations

Ultimately a demonstration that participants know more about specific issues related to coastal transitions (e.g. energy infrastructures or marine policies) is not adequate for appropriately assessing how stakeholders are cognitively, affectively, or behaviourally engaged (Tribbia and Moser, 2008). The application of consultative forms of citizen involvement in coastal transitions—referred to as DAD models—do not meaningfully or effectively engage stakeholders. More involved forms of engagement are required; aligned with the principles of effective stakeholder participation that maximise the inclusion of diverse communities in decision-making processes. We advocate for the MOM method in this chapter as an approach that can overcome the shortfalls of regular limited consultation exercises as purely an exercise in policymaking.

In writing this chapter, we contend that moving towards more meaningful and participatory engagement approaches in coastal transitions will enhance support for, and involvement with, increasing environment and development infrastructures and projects. We see this chapter, not as an attempt to controversially challenge existing dominant and unsustainable involvement practices without recommendations for change, but as a starting point to amend stakeholder engagement methods in line with more effective approaches. Consequently, we outline three areas of focus where future research, policy, and practice should apply the MOM method in design thinking or application.

Future research on stakeholder engagement, and not just in coastal transitions, should clearly identify the extent to which individuals and communities are meaningfully involved with transition processes and ultimate decision-making. We suggest that all research and applied studies provide an "engagement statement" that thoroughly identifies (1) the role of each stakeholder in their current and projected contexts; (2) the relative position of these stakeholders vis-à-vis the environment, technology, resource, or project under question; and (3) the level of influence each stakeholder has on the final decision and a provision of records detailing whether any suggested changes were taken into account. This recommendation allows for transparent engagement processes that can be mapped against Arnstein's (1969) ladder of citizen participation (Figure 3.1). We view this recommendation as a top priority that would shift the balance of power relations away from external positions of industrial-scale governance machines towards decentralised, localised, and bottom-up forms of democratic control. Additionally, further longitudinal analysis, which is often lacking in the domain of socio-ecological assessment, is required to better understand the implications of the MOM method on stakeholders.

Policymakers and policy instruments continue to view stakeholders and communities as a delivery arm for implementing sustainability changes and assume that such groups are inherently supportive. This is predicated on a flawed understanding of the subtleties and specificities of coastal stakeholders and communities, and the diversity of attitudes, cultures, and practices that exist and interact with marine processes. We therefore recommend that state or federal level instituted changes do not provisionally identify, and decide upon, locations for coastal transitions prior to implementing the MOM method. This recommendation, while enhancing the role of coastal stakeholders, allows for reducing changes in policy, public U-turns, as well as minimising wastage of (often) public funds that have been used for costly infrastructure projects that have been inevitably opposed.

There are numerous examples of poor stakeholder engagement through its continued application of consultative forms of participation in public policy decision-making processes. Future applications of stakeholder and public engagement techniques should apply more creative, tailored, and innovative methods to foster meaningful collaborations and support for coastal transitions. To this end, we recommend the inclusion of deliberative democratic techniques such as Delphi panels, citizen assemblies, and the inclusion of coastal communities on executive committees (Devaney et al., 2020; Revez et al., 2020). This ensures that the governance of resources, profits, disruptions, and environments remains, to a large extent, within the control of the stakeholders that historically, culturally, and situationally have relied upon or controlled those environments and resources. By applying these recommendations, we view an improvement in the relationships between various stakeholders whether communities, industry, or policymakers in the continued evolution of sustainability transitions at the coastal zone.

References

Arnstein, S. R. (1969). A ladder of citizen participation. Journal of the American Institute of Planners, *35*, 216–224.

Axon, S. (2019) The human geographies of coastal sustainability transitions, in Heidkamp, C. P. and Morrissey, J. E. (eds.) *Towards Coastal Resilience and Sustainability*, Routledge: Oxon. 276–291.

Berkes, F. (2015) *Coasts for People: Interdisciplinary Approaches to Coastal and Marine Resource Management*. Taylor and Francis: New York, NY.

Boston Globe. (2015) *What Really Toppled Cape Wind's Plans for Nantucket Sound*. https://www.bostonglobe.com/magazine/2015/01/30/what-really-toppled-cape-wind-plans-for-nantucket-sound/mGJnw0PbCdfzZHtITxq1aN/story.html, accessed 11th November 2017.

Carley, S. and Konisky, D. M. (2020) The justice and equity implications of the clean energy transition. *Nature Energy*, **5**: 569–577.

Cash, D. W., Clark, W. C., Alcock, F., Dickson, N. M., Eckley, N., Guston, D. H., Jäger, J., Mitchell, R. B. (2003) Knowledge systems for sustainable development. *PNAS*, **100**: 8086–8091.

Cox, J. R. (2010) Beyond frames: Recovering the strategic in climate communication. *Environmental Communication: A Journal of Nature and Culture*, **4**: 122–133.

Depoe, S. P., Delicath, J. W. and Elsenbeer, M-F. A. (2004) *Communication and Public Participation in Environmental Decision Making*. SUNY Press: Albany, NY.

Devaney, L., Torney, D., Brereton, P. and Coleman, M. (2020) Ireland's citizen assemblies on climate change: Lessons for deliberative public engagement and communication. *Environmental Communication*, **14**: 141–146.

Dietz, T. and Stern, P. C. (eds.). (2008) *Public Participation in Environmental Assessment and Decision Making*. The National Academies Press: Washington, DC.

Few, R., Brown, K. and Tompkins, E. L. (2011) Public participation and climate change adaptation: Avoiding the illusion of inclusion. *Climate Policy*, **7**: 46–59.

Grand Rapids Business Journal. (2010) *Looming Large in the Future*. https://grbj.com/uncategorized/looming-large-in-the-future/, accessed 4th September 2021.

IPCC. (2021) *Climate Change 2021: The Physical Science Basis—Summary for Policymakers*. https://www.ipcc.ch/report/ar6/wg1/downloads/report/IPCC_AR6_WGI_SPM.pdf, accessed 4th September 2021.

Jurjonas, M. and Seekamp, E. (2018) Rural coastal community resilience: Assessing a framework in eastern North Carolina. *Ocean & Coastal Management*, **162**: 137–150.

Jurjonas, M., Seekamp, E., Rivers III, L. and Cutts, B. (2020) Uncovering climate (in)justice with an adaptive capacity assessment: A multiple case study in rural coastal North Carolina. *Land Use Policy*, **94**: 104547.

Kelly, M. R. (2018) Communities, collaboration, and climate adaptation. *Applied Environmental Education and Communication*, **18**: 68–78.

Kelly, C., Ellis, G. and Flannery, W. (2018) Conceptualising change in marine governance: Learning from Transition Management. *Marine Policy*, **95**: 24–35.

McKinley, E. (2022). A Blue Economy for who? Linking marine social sciences with blue economy discourse. Forthcoming lead chapter.

McLachlan, C. (2009) "You don't do a chemistry experiment in your best China": Symbolic interpretations of place and technology in a wave energy case. *Energy Policy*, **37**: 5342–5350.

Morrison, C. and Dearden, A. (2013) Beyond tokenistic participation: Using representation artefacts to enable meaningful public participation in health service design. *Health Policy*, **112**: 179–186.

Poe, M. R., Norma, K. C. and Levin, P. S. (2014) Cultural dimensions of socioecological systems: Key connections and guiding principles for conservation in coastal environments. *Conservation Letters*, **7**: 166–175.

Revez, A., Dunphy, N., Harris, C., Mullaly, G., Lennon, B. and Gaffney, C. (2020) Beyond Forecasting: Using a modified Delphi method to build upon participatory action research in developing principles for a just and inclusive energy transition. *International Journal of Qualitative Methods*, **19**: 1–12.

Susskind, L. and Elliott, M. (1983) *Paternalism, Conflict, and Coproduction: Learning from Citizen Action and Citizen Participation in Western Europe*. Plenum Press: New York, NY.

The Baltimore Sun. (2018) Ocean City's Effort to Keep Windmills Far Offshore Fails as Maryland Delegates Reject Proposal. https://www.baltimoresun.com/news/environment/bs-md-offshore-wind-distance-20180312-story.html, accessed 4th September 2021.

Tribbia, J. and Moser, S. (2008) More than information: What coastal managers need to plan for climate change. *Environmental Science & Policy*, **11**: 315–328.

4

COMMUNITY ACCEPTANCE OF BLUE ENERGY

Identifying Future Research Trajectories for Understanding "Place-Technology-Fit" Perceptions

Stephen Axon

4.1 Blue Energy: Opportunities and Challenges

Offshore renewable energy projects—or so-called Blue Energy installations—are a crucial component of energy policies for many developed, and developing, countries as well as comprising a significant future pathway towards low-carbon electricity globally (Axon, 2019; Devine-Wright and Wiserma, 2020). Though not strictly defined in the literature, Blue Energy is considered energy coming from any form of offshore renewable marine renewable source including offshore wind turbines, floating solar arrays, wave, and tidal installations (Tollefson, 2014; Soma and Haggett, 2015; Nikolaidis et al., 2019). The ocean is regarded as a vast source of renewable and clean energy that exceeds present and projected future energy needs and is expected to play a crucial role in the future energy system (Magagna and Uihlein, 2015; Nikolaidis et al., 2019). Indeed, Blue Energy has featured prominently in many international policies and strategies (e.g. European Commission).

Offshore wind is the fastest-growing activity in the Blue Economy (European Commission, 2017). The European Union is a global leader with around 90% of newly finished offshore wind projects in the world, with the UK, Germany, and Denmark identified as key champions (European Commission, 2017). With 95,471 miles of shoreline, the United States lags significantly behind European offshore renewable energy development. As of 2020, the Block Island offshore wind project, Rhode Island, remained as America's only operational offshore wind energy installation (Russell et al., 2020) with only five offshore turbines. Offshore wind energy installation in the United States remains slow with the Coastal Virginia Offshore Wind pilot project adding just two offshore turbines at the end of 2020. Despite offshore winds typically being stronger and more stable than their onshore-based counterparts, these efficiencies and contributions

DOI: 10.4324/9781003280248-6

to sustainable energy are often muted when taking into consideration existing infrastructures, industries such as shipping and fishing, and community acceptance or opposition (McLachlan, 2009; Russell et al., 2020). These can each be potentially limiting factors. Other forms of Blue Energy face their own distinct challenges and opposition. For example, wave energy installation regularly receives criticism for its potential visual intrusion, negative impacts on marine environments, and possible effects on local (tourist-oriented) economies (Warren et al., 2005; Bailey et al., 2011).

Despite the numerous advantages of installing offshore renewable projects, many proposals treat the coastal zone as a resource (Johnson and Braverman, 2020), and not as a constituent of identity, that often neglects the specific circumstances of coastal places, place identities, and place attachments. Given that sustainability transitions are increasingly integral to coastal areas, it is essential to understand the factors that influence local community acceptance of blue energies (Devine-Wright and Wiserma, 2020), preferably before they are installed (Axon, 2019). This is all the more necessary given the repeated controversies and failures to take into consideration stakeholder perspectives towards the siting of coastal energy projects, e.g. Cape Wind, Massachusetts. Despite the well-established understanding that stakeholder and public engagement is essential to expand upon the understandings of local community acceptance or opposition, decisions regarding development, siting, and installation remain implemented through a top-down approach. This often results in social unacceptability given that communities fail to be "heard" or have their views represented.

This chapter is structured as follows. Section 4.2 provides an overview of community acceptance of Blue Energies, with specific reference to countries in the Global North such as the UK and USA. A key element of public acceptance revolves around "place-technology fit" and is a recurring theme illustrating the need for tailored local and regional understandings should offshore renewable energy installations be proposed. Section 4.3 furthers the rationale for in-depth studies of place-technology fit and community acceptance to be embedded within Blue Energy research, and marine social science more generally. Furthermore, I identify that such research focus on diverse yet meaningful studies that draw upon multiple methodological techniques that enhance our academic understandings, and support practical applications, of Blue Energy installation. Given that this is an under-researched area of investigation, I see these suggestions as potential routes for future research.

4.2 Community Acceptance of Blue Energy and Place-Technology Fit: Integration of Concepts and Practices

Strategic blue growth seeks to attract new technologies and ideas that have a low impact on the marine environment, while also taking into account social interactions necessary for sustainable practices (Soma et al., 2018). As renewable energy infrastructures scale up, it is becoming increasingly common in, and near, where people live (Klain et al., 2017). Consequently, siting these infrastructures has often been controversial and has resulted in project delays and cancellations which can be costly for local governments, developers, and communities. Within this section, I explore the importance of community acceptance of Blue Energies, place-technology fit as a sub-category of acceptance, and indicate how these shape siting preferences with reference to offshore renewable energy development in the UK and USA. Given there are many forms of Blue Energy, I refer to cases of offshore wind energy as examples though I recognise the particularities that may accompany wave, tidal, or floating solar which potentially are not covered in this chapter. While the geographic focus of this chapter corresponds to countries in the Global North, the development of future Blue Energy infrastructures and considerations of community acceptance and place-technology fit are applicable to other development contexts.

4.2.1 Community (Un)acceptance of Blue Energy

It is generally considered good practice and good governance to allow the public to voice their concerns and support—or even rejection—of policies and activities that directly affect them and their communities. Social acceptance is a multi-dimensional concept with inter-related socio-political, market, and community components (Wüstenhagen et al., 2007), despite some empirical research studies that concentrate upon one specific dimension. Despite this, community acceptance refers to the responses of local stakeholders, particularly residents who live close in close proximity to, or in line of sight of, Blue Energy projects (Wüstenhagen et al., 2007; Firestone and Kirk, 2019; Devine-Wright and Wiserma, 2020). Taking into consideration the notion of NIMBYism, many studies have attempted to validate the concept of social acceptance to take into account spatial proximity and siting as determinants of local attitudes (Devine-Wright and Wiserma, 2020). Accounting for public responses is a vital component of project success and timeliness and must be considered when planning Blue Energy installations (Perlaviciute et al., 2018; Axon, 2019). When navigating the process to gain administrative approval for the construction of energy developments (of any kind), meaningful public engagement is essential.

While community acceptance research is dominated by case studies of local responses with primary interest in identifying factors influencing objections or support (Ellis et al., 2007; Firestone et al., 2009), such responses have not yet been meaningfully taken into consideration within decision-making processes (Devine-Wright, 2011). To some extent, this may be explained by the timing of conducting community acceptance research. Relatively few studies have taken place prior to specific projects being proposed (Devine-Wright and Wiserma, 2020). *A priori* studies may have greater impact on policymakers, industry, and civil society since findings can inform pathways of local low-carbon energy development. However, most studies adopt a *post hoc* research design that follows an announcement of a particular proposal and attempts to understand why local communities have adopted particular views or act in certain ways in response.

Though local communities are often sought for their perspectives on proposed projects, this role often takes the form of "tick-box" consulting exercises (Arnstein, 1969; Axon, 2019, 2020). In this purely formal display of public involvement, little action is meaningfully undertaken on behalf of stakeholders to conduct appropriate engagement activities to involve coastal communities. The result is often a missed opportunity for policymakers and developers to meaningfully (re)shape sustainable policy and practice. This lack of engagement is symptomatic of many Blue Energy projects, which reinforce a democratic deficit of public involvement (Peters et al., 2012).

4.2.2 Place-Technology Fit

Place-technology fit is a deeply nuanced concept within community acceptance (McLachlan, 2009; Devine-Wright and Wiserma, 2020) to Blue Energy. Devine-Wright and Wiserma (2020: 2) comments that place-technology fit "refers to the role played by symbolic beliefs about local landscapes or coastscapes, and the degree to which a given technology project is interpreted to be suited or 'out of place' in that context." This is of value given that research has indicated that beliefs about coastscapes and place, conceptualising place both as a locus of meaning and attachment play an important role in shaping community acceptance. Furthermore, community acceptance varies systematically across landscape and coastscape types, with projects considered more acceptable when located in industrialised or military locations and less acceptable in locations regarded as "pristine" or "wild" (Gee, 2010). Partially, this derives from tendencies to "essentialise" symbolic meanings about places and landscapes or coastscapes as "natural" in character, and therefore unsuited to the siting of large-scale energy infrastructure interpreted to "industrialise" or spoil that landscape (Batel and Devine-Wright, 2015). Klain et al. (2017) refer to specific groups of people referred to as "place protectors" who perceive higher value in a specific location due to its impact on environmental, socio-cultural, or economic qualities associated with particular landscapes or coastscapes.

Although effective in summarising complex public sentiments towards energy infrastructure development, NIMBYism fails to explicate the multidimensional nature of complex developments (Devine-Wright, 2012; Devine-Wright and Wiserma, 2020). A thorough criterion that is able to analyse such multidimensional characters is therefore of particular value and application. Place-technology fit is a more sophisticated concept with which to approach community (un) acceptability of energy development (Devine-Wright, 2009). Its first consideration is the geographic makeup of a region; an area's features and publicly held sentiments—considered with respect to place versus space attributes (see Jackson, 2006)—can dictate the (un)acceptance of construction. Secondly, the concept corresponds to technological innovations at the time, which can render a project to "fit"—or not—within the character of a given coastscape for an area. Fit, itself, is the final consideration of the confluence between area value and technology, is ultimately a determinant of project optics (Devine-Wright and Wiserma, 2020).

4.2.3 Integrating Place-Technology-Fit Perception and Preference Studies into Practice

There is substantial value in understanding, and integrating, place-technology-fit perceptions and preferences into practice. Doing so allows for more nuanced understandings of how various publics (e.g. residents, tourists, policymakers, developers, etc.) conceptualise coastscapes and the integration of offshore renewables at "appropriate" locations (McLachlan, 2009; Devine-Wright and Wiserma, 2020). Recommendations outlining conflicts, unintended consequences of development, and how renewables may sit within coastscapes viewed in culturally distinctive ways (Klain et al., 2017) provides policymakers, planners, and developers with accurate viewpoints from local communities that can be integrated *a priori* for greater community input in decision-making processes, particularly where acceptability of siting is considered (Russell et al., 2020). Indicating the inherent conflicts involved with the multi-functional usage of coastal and marine spaces of increasingly placeful environments can help to alleviate the potential for future conflicts over siting particularly. Opposition to offshore renewables that threaten existing activities such as fishing, recreational boating, or tourism can derail projects as exemplified by the Cape Wind Project that defended against more than 20 lawsuits (Boston Globe, 2016). Given the cultural attachments and livelihoods that are intertwined with marine spaces, it is clear that place-technology-fit perceptions and preferences are integral to securing community acceptance.

To avoid the opposition faced by former proposals for Blue Energy, e.g. Cape Wind Project (Kimmell and Stalenhoef, 2011), the concept of place-technology fit is beginning to be integrated into marine social science research (Devine-Wright and Wiserma, 2020). Distinguishing between concepts of space and place, Wiserma (2016) studied cultural attachments to various hypothetically proposed sites for blue energy construction and their relative (un)acceptability in Guernsey (UK). This area of research elucidates the importance of socio-cultural meanings that distinctive coastlines have for various public shape perceptions and preferences (Tollefson, 2014; Wiserma and Devine-Wright, 2014) which ultimately influence acceptance or opposition. Wiserma (2016) explored specific natural and human environments such as "space" being used to explore and enjoy the outdoors, while a "landscape" is useful by means of its visual appeal, e.g. a scenic coastline. A "place," alternatively, is defined as an area that "very much things to be inside of" (Cresswell, 2004: 10). By applying these definitions to search for appropriate placement for offshore blue energy development, the study was able to ascertain responses more acutely. Consequently, there was a trend of unacceptable placements in areas that individuals associated with natural beauty, rendering them as definable areas, and unacceptable for development. This was largely associated with the West coast; which is relatively devoid of human construction (Wiserma, 2016).

Exploring place-technology fit specifically, Wiserma's (2016) participants indicated the (un)acceptability of various blue energy technologies correlated with factors of appearance, size, and location combined. Participants were questioned further about two theoretical wind farms; one with 10–15 turbines to be hypothetically built by 2020, and the other with 20–100 turbines with the latter wind farm representing an all-encompassing utility-scale plant set to be completed in 2025. Participants were more accepting about these developments off the North-West coast, an industrious zone that few were scenically attached to (Wiserma, 2016). Contrasting the accepted notion that residents of a particular coastal area would prefer wind energy development farthest from shore (Knapp and Ladenburg, 2015), Wiserma (2016) found that participants chose locations relatively close to shore, around 6 nautical km from the island. Additionally, tidal developments were identified as being suitable on the North and South-West coastal areas that strengthened the significance of place-technology fit. While tall wind turbines were approved for less scenic areas, different yet similarly visually unappealing locations were chosen for tidal energy development.

The concept of matching a technology to its respective suitable location was found to be acceptable to Guernsey residents, who were able to evaluate the developments' ramifications with geographic site location and subjective interpretation of construction. These understandings clearly indicate the value of integrating place-technology-fit perceptions and preferences for planners and developers for decision-making purposes. Given the opposition to proposals for Blue Energy installations along the US coastline including the Kitty Hawk Project in North Carolina (Energy News Network, 2021), Vineyard Wind in Massachusetts (Time, 2021), and Skipjack Wind Farm in Delaware (The Intercept, 2021).

4.3 Suitable Methodologies and Directions for Future Research in Blue Energy

In this section, I outline a series of thematic areas which warrant further investigation as it corresponds to Blue Energy developments and coastal sustainability transitions. Additionally, I indicate how different methodological designs could illuminate key understandings that support policy objectives and implementation of offshore renewable energy projects, particularly in countries (such as the United States) that have struggled—for various reasons—to develop Blue Energies. It is important to note, however, that methodologies should always be tailored to the objective and scale of the project under investigation. The following five themes are not an exhaustive list of the exact nature and direction of future research on place-technology fit, rather they are suggestions of potential routes to ensuring relevant empirical research is conducted prior to

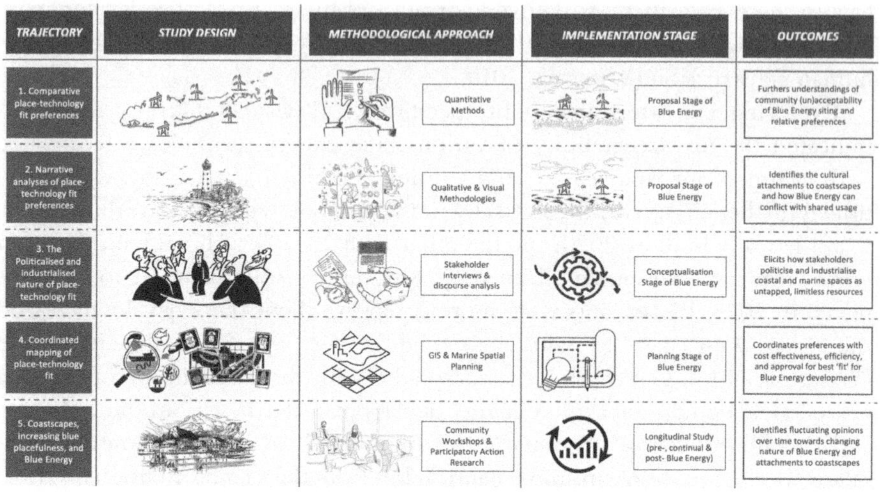

FIGURE 4.1 Research trajectories for exploring place-technology-fit perceptions and preferences

decision-making processes are undertaken without appropriately, or effectively, integrating community responses (see Figure 4.1). Furthermore, it is advantageous for researchers to collaborate with communities in transdisciplinary manners to co-produce findings and identify pathways for sustainability and public engagement (Spreng, 2014; Heaslip and Fahy, 2018) as well as conducting research as part of interdisciplinary research teams.

4.3.1 Residential Comparative Place-Technology-Fit Preferences

Individual, and community, perspectives towards local and distant coastscapes vary depending on place attachments, place identities, and familiarities with particular coastal environments (Devine-Wright and Wiserma, 2020). There is a need to identify the ways in which individuals and communities compare and contrast different coastscapes and the historical developments and cultural meanings that inevitably shape perspectives—including approval and opposition—for Blue Energy developments. Conducting empirical research into this area would illuminate the socio-cultural meanings that are imbued into various coastlines while also indicating which areas may be considered appropriate to install various offshore renewables (Tollefson, 2014; Wiserma and Devine-Wright, 2014). At its centre, this research would identify broad distinctions and preferences of which locations would be acceptable—or strongly opposed—to Blue Energy development (Zichella and Hladik, 2013).

Studies into this thematic area could apply various methodologies. Should this research be conducted quickly, a quantitative approach such as targeted questionnaires in specific communities, a large-scale survey for an entire state, or even a postal survey for residences along the coast would yield results. However, response rates may not be as high in comparison to face-to-face methodologies (Bryman, 2016). The outcomes of this research would provide researchers, policymakers, and developers clear indications (that could be of statistical significance) of locations that are more acceptable to development. Despite this research and involvement method being a clear step to gauge the attitudes and preferences towards Blue Energy siting, it should be acknowledged that this in no way replaces or even compares to true public engagement that meaningfully involves local communities (Axon, 2019).

4.3.2 Narrative Analyses of Place-Technology-Fit Preferences

Understanding the meanings that coastscapes have for those who live on or near them is an essential component to understanding the nature of place-technology fit (McLachlan, 2009; Devine-Wright and Wiserma, 2020). While quantitative measures can provide an overall opinion on a scale that is easily understood by policymakers and developers, true considerations of offshore renewables demand more in-depth analysis. This requires qualitative research techniques that are able to shed light on the ways in which narratives on coastscapes and the integration of Blue Energies are contextualised and in what ways they may be reframed. A study of this nature not only provides the sentiments towards offshore renewables, but also their justification. This provides communities a chance to be "heard," which is often lost when developments are committed to fixed short-term deadlines but are so often the cause of opposition because the community is not provided opportunities to "have their say" (Axon, 2019).

Empirical research conducted under this thematic area of understanding narratives and meanings of place-technology-fit preferences would demand detailed analyses and longer timeframes to undertake qualitative methods appropriately (Creswell, 2009). Focus groups, semi-structured interviews, and visual methodologies such as story maps or pictorial analysis are more adept in exploring the complex and multidimensional sentiments (Conradson, 2005; Mannarini and Fedi, 2009) regarding place-technology-fit preferences than simply reducing these multifaceted narratives and cultural framings to statistical analysis that cannot account for the nuances of support or opposition. This research can therefore be an important first step to explore community acceptance and responsiveness to particular concepts or proposals. Used effectively, this research can be applied within the design and planning processes that may reduce project costs at a later stage when other components have already been settled.

4.3.3 Politicised and Industrialised Variances of Place-Technology Fit

While investigating the responses of residents and communities is essential to offshore renewable developments, there are also significant gaps in understanding of the ways in which Blue Energy—and the Blue Economy at large—is politicised and industrialised at scale by political representatives, industrial stakeholders, non-profit organisations, and think tanks. It should be noted that variances may exist between stakeholders, political parties, actors, and networks that may be supportive, exclusive, or competitive. Consequently, this may have a substantial impact on the ways in which Blue Energy developments are coordinated if mismatches of place-technology fit and overall implementation exist. This could lead to accelerated policy and development proposals as well as acceptances or stagnated processes, falsified bureaucracy, and public resource wastage.

Empirical research under this thematic area should take a more critical lens when evaluating the ways in which Blue Energy and the Blue Economy are politicised, marketed, and industrialised at scale, identifying the implications this has on regional economies and coastal communities. As part of this research stream, I suggest conducting policy analyses on "Blue Plans" or "Blue New Deals" within political discourses and corporate strategies focusing on Blue Growth as part of the business sector (Axon et al., 2022). Combined with stakeholder interviews with key individuals and undertaking analytical frameworks such as discourse analysis would provide such a critical lens to ascertain whether the idealised rhetoric surrounding Blue Energy and the Blue Economy matches the realities and impacts of its implementation.

4.3.4 Mapping Place-Technology-Fit Preferences, Energy Efficiency, Cost-Effectiveness, and Planning Approval Sites

The siting of offshore renewable energy is often contentious, controversial, and subject to appeals and legal challenges (Wiserma and Devine-Wright, 2014). To address this, potential changes to the planning process before decision-making commences could be implemented. One suggestion here relates to incorporating place-technology-fit preferences (e.g. sliding scale positivity ratings), energy efficiency (e.g. kWh generation capacity), cost efficiency (e.g. cost per energy installation), and whether or not particular areas have received planning approval. These data would need to be included within already existing marine spatial planning data and geographic information systems to identify which industries and sectors would be influenced and impacted by Blue Energy development in specific areas.

This thematic stream of research would benefit from spatial research methods but would require the incorporation of results from surveys or interviews around place-technology fit, the energy and cost efficiency variances of offshore renewables by specific location, planning approval sites, and other sectors that are in operation. Applying marine spatial planning and geographic information systems would allow for preferences for Blue Energies—including potentially different types of offshore renewables—to be mapped alongside other components that would aid decision-making processes. The practical application of this research would indicate options as well as trade-offs between efficiencies and public perceptions that would support policymakers and developers in coming to reasoned decisions for siting in advance. This thematic area of research could easily be integrated with others as part of a wider community engagement strategy.

4.3.5 Reconceptualising Coastscapes as a Result of Increasing Blue Placefulness and the Implications for Offshore Renewables

With increasing attention paid to the Blue Economy, and with future Blue Energies becoming increasingly likely, there is a need for empirical research to document longitudinally how perspectives towards coastscapes may fluctuate over time with increasing placefulness and changing notions of development along coastlines. Given the global increase of Blue Economy innovations, it is essential that coastal communities not only have ample opportunities to provide their opinions on changes to their local areas but also indicate the positive, and negative, impacts of the Blue Economy. It is inevitable that changes to coastal communities will occur over time. Despite these changes, the coast provides important cultural and economic ecosystem services that coastal communities depend upon. The implications of increasing development in these areas may also alter the socio-cultural meanings of the coast. Such research is necessary as both coastal environments, technological developments, and social attitudes fluctuate over time and are not static. It should not be taken for granted therefore that initial approval for offshore renewable energy projects remain or continue to be viewed positively should subsequent environmental changes, policy directions, or events occur.

Continual, and longitudinal, research is integral to providing holistic understandings of community acceptance of Blue Energy and the Blue Economy overall. To this end, I suggest that mixed methods research allow for both short "snapshot" understandings of public perspectives towards the ways in which coastsscapes have altered and a detailed comprehension of the socio-economic impacts developments have had on coastal communities. Research conducted as part of this thematic area could apply quantitative approaches such as annual surveys to identify attitudinal changes and qualitative techniques such as focus groups and/or community workshops that provide opportunities to justify how sentiments have enhanced, weakened, or changed towards coastscapes and offshore renewables. Such research is useful to policymakers and developers to ensure effective community engagement strategies are able to be proactive in mitigating negative issues and adapting to changes in social attitudes. This research would be of particular importance to coastal areas that have established Blue Energies and wish to integrate additional innovations as they correspond to the Blue Economy.

4.4 Conclusions and Recommendations

Considering the dangers, the planet will face as the consequences of the climate crisis continue to be exacerbated by human activity, coastal communities are facing significant threats to their properties, shorelines, industries, and energy generation sectors. To enhance climate mitigation and enhance resiliency, it is necessary to reduce energy-production-related emissions and enhance the development of offshore renewable energy systems. This requires transitions to renewable—including blue—energy solutions. This shift requires large-scale changes in infrastructure, as new power plants, transmission lines, and infrastructure will be required. However, the impacts that public (un)acceptance can have on proposed blue energy projects ensures that future development has public support. The concepts of place-technology fit and NIMBYism can inform developers as to where blue energy development would be most positively—and negatively—supported, as well as how the public may respond to construction proposals. Thus, an effective transition to Blue Energies requires effective public engagement in the development process that takes into consideration factors that shape place-technology-fit preferences. Engaging coastal communities with their preferred options for development is deemed to be critical for the success of Blue Energy (Axon, 2019). The five thematic areas and associated methodologies outlined in this chapter provide potential trajectories for future marine social science research to ensure that such a transition comprising Blue Energy follows effective public engagement principles and practices as well as inclusive of the findings of such methodologies into decision-making processes. Further, I recommend that researching, and potentially enhancing, community acceptance of Blue Energy installations is done as part of an interdisciplinary and transdisciplinary approach (Spreng, 2014; Heaslip and Fahy, 2018).

In this regard, I hope that this chapter provides a useful starting point rather than an endpoint to studies exploring Blue Energy developments. Particularly, I call for increasing attention to be paid to coastscapes, their meaning for residents, tourists, and policymakers, as well as the implications of offshore renewable energy implementation may have for the natural, human, and built environments. This requires substantive empirical research in order for policy analyses and practical applications to be sensitive and responsive to attitudes towards coastal developments that may be better explained by underlying place-technology-fit perceptions. It would be unacceptable for local, regional, or national authorities and organisations to undertake developments without understanding the local contexts and meanings that coastscapes provide. Should policymakers and developers follow, or incorporate, our suggestions it would not be unreasonable to assume a greater involvement of community perspectives that may reduce opposition to offshore renewables later on during a project timescale.

References

Arnstein, S. R. (1969). A ladder of citizen participation. Journal of the American Institute of Planners, 35, 216–224.

Axon, S. (2019). The human geographies of coastal sustainability transitions, in Heidkamp, C. P. and Morrissey, J. E. (eds.) Towards Coastal Resilience and Sustainability, Routledge: Oxon, 276–291.

Axon, S. (2020). The socio-cultural dimensions of community-based sustainability: Implications for transformational change. Journal of Cleaner Production, 266, 121933.

Axon, S., Bertana, A., Graziano, M., Cross, E., Smith, A., Axon, K. and Wakefield, A. (2022). A U.S. Blue New Deal: What does it mean for just transitions, sustainability, and resilience of the Blue Economy? The Geographic Journal. https://doi.org/10.1111/geoj.12434

Bailey, I., West, J. and Whitehead, I. (2011). Out of sight but not out of mind? Public perceptions of wave energy. Journal of Environmental Policy & Planning, 13 (2), 139–157.

Batel, S. and Devine-Wright, P. (2015). A critical and empirical analysis of the national-local 'gap' in public responses to large-scale infrastructures. Journal of Environmental Planning and Management, 58, 1076–1095.

Boston Globe. (2016). Offshore Wind May Finally Take Off with Big Projects, None Named Cape Wind. https://www.bostonglobe.com/magazine/2016/03/23/offshore-wind-may-finally-take-off-with-big-projects-none-named-cape-wind/FZ9Ng715HYkgKFFoNtlZHN/story.html, accessed 11th November 2021.

Bryman, A. (2016). Social Research Methods, Oxford University Press: Oxford.

Conradson, D. (2005). Focus groups, in Flowerdew, R. and Martin, D. (eds.) Methods in Human Geography, Pearson: Essex.

Creswell, J. W. (2009). Research design: Qualitative, Quantitative and Mixed Methods Approaches, SAGE Publications Ltd: London.

Cresswell, T. (2004). Place: A Short Introduction, Blackwell: Oxford.

Devine-Wright, P. (2011) Rethinking NIMBYism: The role of place attachment and place identity in explaining place-protective action. Journal of Community & Applied Social Psychology, 19, 426–441.

Devine-Wright, P. (2011). Public engagement with large-scale renewable energy: Breaking the NIMBY cycle. WIRES Climate Change, 2, 19–26.

Devine-Wright, P. (2012). Explaining "NIMBY" objections to a power line: The role of personal, place attachment and project-related factors. Environment and Behavior, 45, 761–781.

Devine-Wright, P. and Wiserma, B. (2020). Understanding community acceptance of a potential offshore wind energy project locations: An island-based analysis of 'place-technology fit'. Energy Policy, 137, 111086.

Ellis, G., Barry, J. and Robinson, C. (2007). Many ways to say no, different ways of saying yes: Applying Q-methodology to understand public acceptance of wind farm proposals. Journal of Environmental Planning and Management, 50, 517–551.

Energy News Network. (2021). N.C. Offshore Wind Faces Big Challenges, Even Bigger Opportunities, Advocates Say. https://energynews.us/2021/09/30/n-c-offshore-wind-faces-big-challenges-even-bigger-opportunities-advocates-say/, accessed 10th December 2021.

European Commission. (2017). Report on the Blue Growth Strategy: Towards More Sustainable Growth and Jobs in the Blue Economy, European Commission: Brussels.

Firestone, J. and Kirk, H. (2019). A strong relative preference for wind turbines in the United States among those who live near them. Nature Energy, 4, 311–320.

Firestone, J., Kempton, W. and Kreuger, A. (2009). Public acceptance of offshore wind projects in the USA. Energy Policy, 35: 1584–1598.

Gee, K. (2010). Offshore wind power development as affected by seascape values on the German North Sea coast. Land Use Policy, 27, 185–194.

Heaslip, E. and Fahy, F. (2018). Developing transdisciplinary approaches to community energy transitions: An island case study. Energy Research & Social Science, 45, 153–163.

Jackson, P. (2006). Thinking geographically. Geography, 91 (3), 199–204.
Johnson, E. R. and Braverman, I. (2020). Blue legalities: Governing more-than-human oceans, in Braverman, I. and Johnson, E. R. (eds.) Blue Legalities: The Life & Laws of the Sea. Duke University Press: Durham, NC, USA.
Kimmell, K. & Stalenhoef, D. S. (2011). The Cape Wind offshore wind energy project: A case study of the difficult transition to renewable energy. Golden Gate University Environmental Law Journal, 5, 197–225.
Klain, S. C., Satterfield, T., MacDonald, S., Battista, N. and Chan, K. M. A. (2017). Will communities "open up" to offshore wind? Lessons learned from New England islands in the United States. Energy Research & Social Science, 34, 13–26.
Knapp, L. and Ladenburg, J. (2015). How spatial relationships influence economic preferences for wind power—A review. Energies, 8, 6177–6201.
Magagna, D. and Uihlein, A. (2015). Ocean energy development in Europe: Current status and future perspectives. International Journal of Marine Energy, 11, 84–104.
Mannarini, T. and Fedi, A. (2009). The quality of participation in the perception of citizens: Findings from a qualitative study, in Lange, N. and Wagner, M. (eds.) Community Psychology: New Developments, Nova Science Publishers: New York.
McLachlan, C. (2009). "You don't do a chemistry experiment in your best China": Symbolic interpretations of place and technology in a wave energy case. Energy Policy, 37 (12), 5342–5350.
Nikolaidis, G., Karaolia, A., Matsikaris, A., Nikolaidis, A., Nicolaides, M. and Georgiou, G. C. (2019). Blue Energy potential analysis in the Mediterranean. Frontiers in Energy Research, 7, 62.
Perlaviciute, G., Schuitema, G., Devine-Wright, P. and Ram, B. (2018). At the heart of a sustainable energy transition: The public acceptability of energy projects. IEEE, 16, 49–55.
Peters, M., Fudge, S., Hoffman, S. and High-Pippert, A. (2012). Carbon management, local governance and community engagement. Carbon Management, 3, 357–368.
Russell, A., Firestone, J., Bidwell, D. and Gardner, M. (2020). Place meaning and consistency with offshore wind: An island and coastal tale. Renewable and Sustainable Energy Reviews, 132, 110044.
Soma, K. and Haggett, C. (2015). Enhancing social acceptance in marine governance in Europe. Ocean and Coastal Management, 117, 61–69.
Soma, K., van den Burg, S., Hoefnagel, E., Stiuver, M. and van der Heide, M. (2018). Social innovation: A future pathway for Blue Growth. Marine Policy, 87, 363–370.
Spreng, D. (2014). Transdisciplinary energy research—Reflecting the context. Energy Research & Social Science, 1, 65–73.
The Intercept. (2021). Oil-Backed Group Opposes Offshore Wind—For Environmental Reasons. https://theintercept.com/2021/12/08/oil-industry-wind-farm-prevent/, accessed 10th December 2021.
Time. (2021). U.S. Fishermen Are Making Their Last Stand against Offshire Wind. https://time.com/6102900/offshore-wind-fishing/, accessed 10th December 2021.
Tollefson, J. (2014). Power from the oceans: Blue Energy. Nature, 508, 302–304.
Warren, C. R., Lumsden, C., O'Dowd, S. and Birnie, R. V. (2005). 'Green on Green': Public perceptions to wind power in Scotland and Ireland. Journal of Environmental Planning and Management, 48 (6), 853–875.
Wiserma, B. (2016). Public acceptability of offshore renewable energy in Guernsey: Using visual methods to investigate local energy deliberations, Unpublished PhD Thesis, University of Exeter, Exeter, UK.

Wiserma, B. and Devine-Wright, P. (2014). Public engagement with offshore renewable energy: A critical review. WIRES Climate Change, 5 (4), 493–507.
Wüstenhagen, R., Wolsink, M. and Bürer. M. J. (2007). Social acceptance of renewable energy innovation: An introduction to the concept. Energy Policy, 35, 2683–2691.
Zichella, C. and Hladik, J. (2013). Siting: Finding a home for renewable energy and transmission. The Electricity Journal, 26 (8), 125–138.

5

CONFLICTS AND COMMUNITIES

Marine Aquaculture and the Blue Economy

Teresa R. Johnson and Samuel P. Hanes

5.1 Introduction

Marine social science has a more central role to play in producing knowledge necessary to advance the Blue Economy agenda sustainably and inclusively. Marine aquaculture is one sector of the Blue Economy where marine social science research is especially important. As the fastest-growing food production system in the world, aquaculture is considered a source of great economic potential for coastal regions. However, marine aquaculture can conflict with traditional uses and values, resulting in conflicts in some, but not all, places. When conflicts do emerge, they can significantly impede the development of marine aquaculture (Hanes, 2018). Key concerns underlying marine aquaculture conflicts involve environmental impacts and displacement and disruption of traditional fishing practices and property rights systems (Walters, 2007; Knapp and Rubino, 2016; Hanes, 2018). As well, speed and scale of aquaculture development, demographic trends, and socioeconomic context influence aquaculture conflicts (Hanes, 2018). In many cases, subsets within communities are reluctant to support aquaculture when they perceive that they will disproportionately experience negative social and environmental impacts (Gibbs, 2009), including threats to their environmental beliefs, values, and landscape preferences (Hanes, 2018). Given the fact that social factors and conflicts are the most critical impediments to aquaculture development (Gentry et al., 2017; Weitzman and Filgueira, 2020), marine social science focused on understanding local marine aquaculture conflicts and acceptance, and the complex values that underlie them, is critical for advancing this part of the Blue Economy.

The concept of a carrying capacity recognises limits to growth and has been applied to the development of aquaculture (e.g. Weitzman and Filgueira, 2020). Social carrying capacity is considered to be the most important determinant

DOI: 10.4324/9781003280248-7

of aquaculture expansion (e.g. Weitzman and Filgueira, 2020). It has been defined as the level of farm development that causes unacceptable social impacts (McKindsey et al., 2006) or that the community is willing to allow (Gibbs, 2009), although the concept still requires further defining and exploration (Weitzman and Filgueira, 2020). Typically inferred from measures of social acceptance and conflicts, communities exhibiting low social acceptance or high conflict levels over existing or proposed farms indicate that aquaculture has reached or surpassed the point the community is willing to allow (Gibbs, 2009). Social carrying capacity ultimately reflects a limit being set on one ocean group's use and access to the Blue Economy. Due to its socially constructed nature, understanding stakeholder perceptions, identity, and values are fundamental to any social carrying capacity assessment and marine social scientists can gather such understanding in many ways, including focus groups, surveys, and interviews (Gibbs, 2009), and through analyses of existing data, like public testimony (Hanes, 2018) and the media (Kluger et al., 2019). Scholars have also incorporated visual aspects of aquaculture in their assessments of social carrying capacity via mail surveys (Dalton et al., 2017). Social carrying capacity of aquaculture must be understood within the context of the broader social-ecological system in which it is situated, necessitating the integration of multiple methods, disciplines, and perspectives (Johnson et al., 2019). This is echoed in Krause et al.'s (2015, pg. 53) call for "creative combinations of theories and methods widely applicable to assess and interpret the social dimensions of aquaculture." Marine social science—encompassing of a wide range of social science disciplines—is well positioned to answer a diverse set of fundamental questions about the human dimensions of the Blue Economy, including aquaculture (McKinley, Chapter 2, this volume).

To this growing body of marine social science research on marine aquaculture and the Blue Economy, this study contributes a comparative analysis of conflicts and social acceptance in three aquaculture regions in Maine (USA). Our study drew from multiple marine social science perspectives and combined quantitative assessments of conflicts and social acceptance, while being grounded in qualitative, ethnographic research. Although we hypothesised that there would be relatively low social carrying capacities in all three study regions, results revealed more complex and nuanced aspects of social carrying capacity of marine aquaculture, underscoring the opportunity and need to integrate marine social science into Blue Economy research agendas (McKinley, Chapter 2, this volume).

5.2 Research Setting

This research occurred in the state of Maine, which has both the most rural and oldest population in the United States. Along with a decline in abundance or access to traditional wild fisheries, parts of Maine are experiencing post-industrial restructuring that includes amenity migration (i.e. in-migrants drawn to environmental amenities) (Hanes, 2018). Maine's fishery-dependent communities are considered to be overly reliant on a single commercial fishery, American lobster, which represents nearly 80% of all revenue generated by commercial fisheries in the state (Johnson, 2020). Marine aquaculture is seen as an opportunity to diversify working waterfronts, and thus its role in the state's Blue Economy is expected to grow (Johnson, 2020).

Indeed, aquaculture in Maine has grown considerably in the last 10 years (ME DMR, 2018), with continued growth expected following strategic investments. The most important aquaculture crops in Maine are currently salmon, oysters, and mussels (ME DMR, 2018). All aquaculture production is subject to the state's formal governance process that establishes rights to individuals to lease areas for aquaculture. Since 1973 Maine Aquaculture Law has allowed individuals to lease areas for aquaculture, with the Maine Department of Marine Resources (ME DMR) having authority over siting aquaculture farms. The state offers three leasing options for farmers: Standard leases are granted for up to 100 acres for 20 years maximum; and they can be used to grow both bottom and suspended culture and shellfish, finfish, and/or marine algae. Experimental Leases are designed to allow for commercial and scientific research on sites before a longer or larger lease is granted, are much smaller, up to 4 acres maximum, and granted for only three years, with no renewal allowed except for scientific research. Since 1999, limited-purpose aquaculture licenses (LPAs) have been granted for much smaller areas, up to 400 square feet, for only one year at a time, but with annual renewals allowed. LPAs can be used to grow only shellfish and seaweeds, not finfish, and are intended for farmers testing new sites and for small commercial enterprises and hobby farmers. LPAs do not require public hearings and are less expensive than standard leases. The number of LPAs has increased from about 200–600 over the past three years. This combination, aquaculture growth on a coast with both declining populations and growing, gentrifying regions, makes Maine an ideal place to examine factors affecting social carrying capacity of marine aquaculture.

5.3 Methods

Authors embraced a mixed-methods research approach (Creswell and Clark, 2014) that combined quantitative assessments of conflicts and social acceptance with ethnographic research to compare social carrying capacity in three marine aquaculture regions in Maine: Damariscotta, Blue Hill, and Downeast (Figure 5.1). Damariscotta and Blue Hill are characterised by a more affluent citizenry that is less dependent upon natural resources compared to the population found in Downeast (Table 5.1). Downeast is further distinguished by hosting the majority of the state's finfish (salmon) farms, while oysters dominate aquaculture production in Damariscotta and Blue Hill. Table 5.1 summarises socioeconomic data from the US Census Bureau's American Community Survey (2017) and available aquaculture lease and license data from the Maine Department of

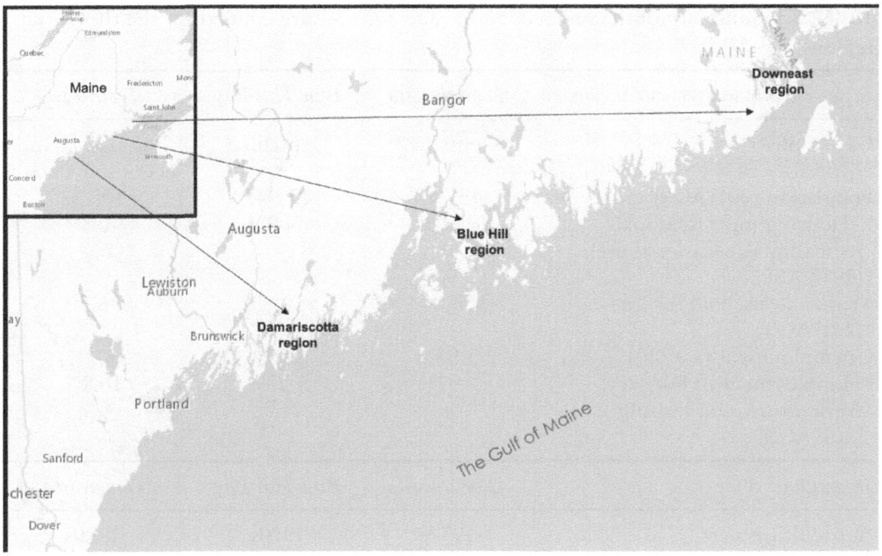

FIGURE 5.1 Map depicting three study regions: Damariscotta, Blue Hill, and Downeast

Marine Resources[1] for the three regions. Based on this socioeconomic context, we hypothesised that there would be relatively low social carrying capacities in all these regions due to (1) the predominance of finfish aquaculture in Downeast and (2) the more gentrified (wealthy) communities in Damariscotta and Blue Hill.

The first author has been engaged in ongoing ethnographic research and stakeholder engagement with aquaculture industry and other stakeholders since 2012, with the Aquaculture in Shared Waters project, a training program for fishers interested in adopting seaweed and shellfish aquaculture (Cleaver et al., 2018). That work continued as part of a transdisciplinary research project known as the Sustainable Ecological Aquaculture Network (SEANET), a $20 million project funded by the US National Science Foundation, that examined marine aquaculture in Maine from a social-ecological systems perspective (Johnson et al., 2019; Bricknell et al., 2021). Results from over 50 key informant interviews conducted between 2015 and 2016 (e.g. Hanes, 2018; Johnson and Hanes, 2018; Johnson et al., 2019) and this ongoing ethnographic research provided insight into stakeholders' values and preferences and understanding about the regional geographic and historical context of Maine's coastal communities where marine aquaculture is developing. Statements used in the structured survey, described below, were informed by this ethnographic research and analysis of stakeholder interviews.

To provide a quantitative measure of conflicts, an indicator of social carrying capacity, we updated and reanalysed an existing dataset consisting of all available lease hearing transcripts from the ME DMR (Hanes, 2018; Johnson and Hanes, 2018). The final dataset analysed in this paper includes 330 lease hearing

TABLE 5.1 Summary of the socioeconomic and aquaculture context for the three study regions

Socioeconomic and geographic context	*Damariscotta*	*Blue Hill Bay*	*Downeast*
Geography	Gentrified	Gentrified	Population and economic decline
Population (2017ACS)	15,286	12,847	10,127
% Housing units (seasonal/ recreational/occasional use) (2017 ASC)	34.2%	34.0%	19%
Median household income (2017ACS)	$58,064	$50,852	$36,306
Unemployment rate (2017ACS)	3.58%	6.22%	8.7%
% Employment in Forestry, Agriculture, and Fishing (2017ACS)	4.97%	5.28%	10.0%
Aquaculture context	*Damariscotta*	*Blue Hill Bay*	*Downeast*
Aquaculture start	1970s	1970s	1980s
Acreage	147.8	103.7	517.3
#Farms	85	38	36
• #LPAs	50	25	16
• #Standard Farms	31	11	20
• #Experimental	4	2	0
Primary type of AQ today	Shellfish	Shellfish	Finfish

transcripts (Table 5.2). This dataset represents the public lease hearing history for all current and recent aquaculture leases in Maine, i.e. for all current aquaculture leases and those terminated after 2013. All lease hearings were entered into a QSR NVivo 11 database for qualitative analysis. For each hearing transcript, concerns raised were coded and quantified. In addition, each hearing was coded as one of three types: Conflict, Uncertainty, or No Concerns. Conflict hearings refer to those where 10 or more concerns were raised and characterised by conflicts. Indicators of these were the presence of organised opposition, often represented by legal teams. Uncertainty hearings represent hearings where a few concerns or questions were raised without organised opposition and the hearing discussion was mostly informational. The hearing transcripts also provide qualitative information about the regional context of aquaculture development in each region related to stakeholders' values, conflicts, and acceptance.

TABLE 5.2 Summary of lease hearing conflicts analysis

	Damariscotta	*Blue Hill*	*Downeast*	*Rest of Maine*
Total no. of lease hearings	69	30	87	144
• # Hearings with no concerns	44	13	70	93
• # Conflict hearings	2	8	0	8
• # Uncertainty hearings	23	9	17	43
% Hearings in the state	21%	9%	26%	44%
Total no. of concerns	87	283	46	281

We also administered a mail survey to a random sample of 2000 residents in 2018; 500 each were sent to Damariscotta and Blue Hill regions, and 1000 were sent to the Downeast region. A marketing company provided the random sample. Following a modified Dillman design (Dillman et al., 2014), an introductory letter was mailed for each survey, followed by two rounds of the survey, with the subsequent mailings only sent to non-respondents. In addition to collecting data on respondents' geographical context, experience with aquaculture, and socioeconomic background, we also collected data to assess social acceptance and social conflicts by asking them to reflect on areas of conflict that had emerged in prior research (e.g. Hanes, 2018; Johnson and Hanes, 2018).

First, we asked participants to share their opinion regarding the amount of aquaculture [Far too much, Too much, Just the right, Too little, Far too little] and the speed of aquaculture development [Far too fast, Too fast, Just right speed, Too Slow, Far too slow] in their area. Responses were combined to form a single variable (Growth), and data were transformed so that higher mean values reflect views that growth is at or beyond their satisfaction (i.e. at or beyond social carrying capacity) and lower mean values reflect desire for more growth. Second, we asked participants to report their general view about marine aquaculture on a scale from Very Negative to Very Positive and their level of satisfaction with the governance of marine aquaculture [Very Unsatisfied to Very Satisfied], both on a 6-point Likert scale. Third, we asked participants to report their level of agreement on a 6-point Likert scale with thirteen statements reflecting key dimensions of marine aquaculture.

Because normality could not be assumed, Kruskal-Wallis H nonparametric tests were conducted to test for significant differences between regions in their views about aquaculture, with $p = .05$. Where regional significant differences were observed, means and standard error of the mean were calculated for each region for growth, general views of aquaculture, satisfaction with governance, and the aquaculture values (Figure 5.2). Post-hoc Mann-Whitney tests were then conducted when significant differences were observed, using the Bonferroni correction, $p = .017$ (Table 5.3).

5.4 Results

We first briefly summarise relevant aspects of the geographic socioeconomic context of aquaculture in these three regions that were informed by interviews and ethnographic research in the regions and then present findings from the conflict analysis and mail survey.

5.4.1 Regional Aquaculture Context

The birth of Maine's aquaculture industry occurred in the gentrified region of Damariscotta, where the first lease in the state was granted in 1973. Nearly 66% of all oysters produced in Maine in 2018 came from this area (ME DMR, 2018).

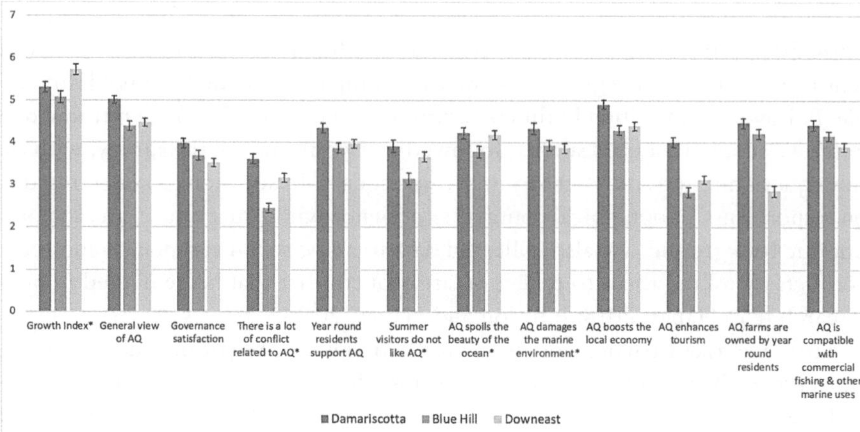

FIGURE 5.2 Mean responses for views about the growth of aquaculture, general views and satisfaction with governance, and value dimensions of aquaculture where regional differences were observed

A local university professor used a cooperative extension research approach to lead a group of innovators who have made Damariscotta the centre of oyster aquaculture in the State. A locally based, Maine Sea Grant and Cooperative Extension team member has helped connect the University, the industry, and the community. A local oyster festival illustrates the positive relationships that have formed between them and the community.

The gentrified region surrounding Blue Hill has a more contentious history related to aquaculture. Located near Acadia National Park, it has long drawn residents seeking the park's picturesque beauty while avoiding the crowds found in towns adjacent to the park (Woodward, 2015). The situation has been described as "NIMBY-like" (Hanes, 2018). To avoid conflicts at the lease hearings, farmers have opted for LPA licenses because they do not require public hearings; 66% of the farms in the region are these very small farms (Table 5.1). Conflict related to aquaculture first emerged in Blue Hill in 1999 when a salmon farm was proposed; this led to the formation of the Friends of Blue Hill Bay, an organisation that successfully opposed the proposal and other subsequent proposals (Hanes, 2018).

In contrast, Downeast remains quite rural and in an early stage of post-industrial transition. Once a centre of the US sardine canning industry, this is the only place in Maine with finfish farms today. The first finfish farm was established in 1982 in this region, increasing to 13 by 1990. However, the industry experienced rapid consolidation and by 2004 a single company had purchased all of the commercial operations in Maine. Today it remains the sole commercial finfish farming company in the state, with 24 lease sites employing local residents. Like Damariscotta, the region has benefited from the work of a locally based, Maine Sea Grant and Cooperative Extension team member who connects the University, the industry, and the community.

TABLE 5.3 Survey statements used to investigate social carrying capacity (SCC) and acceptance, and key value and policy dimensions of aquaculture (AQ) in the three study regions. Results from Kruskal-Wallis H test ($p = .05$) and directions of significant differences when determined via post-hoc Mann-Whitney U tests ($p < .017$)

	Kruskal-Wallis H	*K-W Asymp. Sig.*	*Direction of significant differences (Mann–Whitney, p < .017)*
SCC/Acceptance statements			
Growth level	15.962	0	DE > Dam (.005); DE > BH (.000)
General view of AQ	19.552	0	Dam > BH (.001); Dam > DE (.000)
Satisfaction with AQ governance	10.798	0.005	Dam > DE (.000)
Value statements			
There is a lot of conflict related to AQ	47.061	0	BH > Dam (.001); DE > Dam (.006); BH > DE (.000)
Summer visitors do not like AQ	15.63	0	BH > Dam (.001); BH > DE (.004)
Year-round residents support AQ	10.576	0.005	Dam > BH (.002); Dam > DE (.009)
AQ is compatible with commercial fishing and other marine uses	13.199	0.001	Dam > DE (.000)
AQ boosts the local economy	17.161	0	Dam > BH (.000); Dam > DE (.001)
AQ enhances tourism	46.948	0	Dam > BH (.000); Dam > DE (.000)
AQ farms are owned by year-round residents	75.277	0	Dam > DE (.000); BH > DE (.000)
AQ damages the marine environment	8.945	0.011	BH > Dam (.014); DE > Dam (.005)
AQ spoils the beauty of the ocean	7.22	0.027	–
AQ is important to the way of life	5.432	0.066	–
AQ creates navigational problems	4.31	0.116	–
AQ is more sustainable	4.179	0.124	–
AQ provides good jobs	4.081	0.130	–
Policy statements			
AQ farmers should complete training before they are able to farm in Maine's ocean	10.687	0.005	BH > Dam (.002); BH > DE (.013)
The government should monitor all AQ farms regardless of their size or use	12.271	0.002	BH > Dam (.000)
All proposed AQ farms should require public hearings regardless of their size or use	12.897	0.002	BH > Dam (.000)
The government should consider visual impacts before approving an AQ farm	7.497	0.024	BH > Dam (.012)
There should be a limit on the number of AQ farms an individual/company can have	10.719	0.005	BH > DE (.001)
There should be a limit on the total space that AQ can take up in an area	8.157	0.017	BH > DE (.009)
Land owners should decide if a farm is allowed in an area	4.582	0.104	–

5.4.2 Conflicts Analysis

Results from the analysis of 330 lease hearing transcripts in Maine for the three regions revealed (1) spatial variation in conflict across the three regions, with greater conflict observed in Blue Hill compared to the other two regions, and (2) a relatively low level of conflict across the dataset (Table 5.2). Of the 330 lease hearing transcripts analysed across the state, 186 (56%) were located in the three study regions. Of the three study regions, Downeast had the most hearings (87), followed by Damariscotta (69) and Blue Hill (30). Most hearings across the state (67%) and in each region except Blue Hill were characterised as having No Concerns. Concerns were raised at only 110 hearings across the state (33%), with only 18 (5.4%) of these being characterised as Conflict Hearings. Nearly half of all the Conflict Hearings across the state occurred in the Blue Hill region (44%), despite the region only having 9% of the total hearings. There were zero Conflict Hearings in Downeast and only two in Damariscotta. Five of the eight Conflict Hearings occurring outside the study region were located in areas adjacent to Blue Hill. Instead, Downeast and Damariscotta were characterised higher number of Uncertainty hearings (27% and 33%, respectively) or hearings with no concerns (80% and 64%, respectively). Blue Hill hearings were nearly equally split between having no concerns, conflicts, and uncertainty hearings (43%, 27%, and 30%, respectively). Across the state, 41% of the 697 concerns raised at aquaculture lease hearings occurred in Blue Hill (Table 5.2).

5.4.3 Mail Survey

Of the 2000 surveys mailed, 274 were undeliverable and 440 were returned, for an adjusted response rate of 25.6%. Of the 440 returns, 123 were from Damariscotta, 135 from Blue Hill, and 182 from Downeast. The survey results were biased slightly towards males (60.2%) compared to females (38%). Participants ranged from 22 to 92 years, with a mean age of 62.61 years old, which is consistent with the state's median age (44.3 years). Similarly consistent with Maine's population, 49.5% of participants reported incomes over $50,000, with median income in the state being about $53K, and 96.4% of respondents reported having a high school diploma or equivalent, which is comparable to the state (92.1%).

Preference for more aquaculture growth was highest in the Downeast region (Figure 5.2). The three regions differed significantly in their views about aquaculture growth (Kruskal-Wallis, $p = .000$), with scores in the Downeast region reflecting significantly greater interest in more growth of aquaculture compared to the Damariscotta (Mann-Whitney, $p = .005$) and the Blue Hill (Mann-Whitney, $p = .000$) regions (Table 5.3).

Views about the marine aquaculture industry were more positive in the Damariscotta compared to Blue Hill and Downeast (Figure 5.1). The three regions differed significantly in terms of residents' general views about aquaculture (Kruskal-Wallis, $p = .000$), with Damariscotta having significantly more positive views than the residents in Blue Hill (Mann–Whitney, p =.000) and Downeast (Mann-Whitney, p = .000) (Table 5.3). Similarly, satisfaction with aquaculture governance was highest in Damariscotta (Figure 5.1); the three regions differed significantly in their satisfaction with aquaculture governance (Kruskal-Wallis, p = .005), with significantly higher levels of satisfaction seen in Damariscotta compared to the Downeast region (Mann-Whitney, p =.001) (Table 5.3).

The three regions differed significantly on nine of the thirteen aquaculture value dimensions (Table 5.3), with mean responses reflecting more positive views about aquaculture in Damariscotta compared to the other two regions (Figure 5.2). Significantly more Blue Hill residents believed there to be a lot of aquaculture conflicts compared to Damariscotta and Downeast residents (Mann-Whitney, p = .000), while significantly more Downeast residents believed the same compared to Damariscotta residents (Mann-Whitney, p = .006). Significantly more Blue Hill residents viewed summer visitors as not liking aquaculture compared to Damariscotta residents (Mann-Whitney, p = .000) and Downeast residents (Mann-Whitney, p = .004). Significantly more Damariscotta residents believe year-round residents support aquaculture than Blue Hill residents (Mann-Whitney, p = .002) and Downeast residents (Mann–Whitney, p = .009). Significantly more Damariscotta residents view aquaculture as compatible with commercial fishing and other marine uses more than residents in the Downeast region (Mann–Whitney, p = .000). Significantly more Damariscotta residents viewed aquaculture as boosting the local economy and enhancing tourism than residents in Blue Hill and the Downeast region (Mann-Whitney, $p < .001$). Significantly fewer Downeast residents viewed aquaculture farms as being owned by year-round residents compared to those in Damariscotta and Blue Hill (Mann-Whitney, p = .000). Significantly fewer Damariscotta residents viewed aquaculture as damaging the marine environment compared to residents in Blue Hill (Mann-Whitney, p = .014) and Downeast (Mann-Whitney, p = .005).

5.5 Discussion

There is urgent need to allow society to engage in Blue Economy dialogue, necessitating a more complex and nuanced understanding of the interactions between people and the sea, including non-economic understanding (McKinley, Chapter 2, this volume). As the world's fastest-growing food production system, marine aquaculture will undoubtedly continue to be a critical component of the Blue Economy. Social conflicts, often arising from disparate values and interests, are critical impediments to marine aquaculture development. Conflicts infer the existence of a limit of aquaculture growth that society is willing to accept, with the greater social acceptance and less conflict being indicative of a place with a higher (or not yet met) social carrying capacity. This marine social science study on social carrying capacity of marine aquaculture combined quantitative assessments of conflicts and social acceptance with qualitative, ethnographic research to reveal three frames associated with marine aquaculture development and in doing so underscores the importance of considering the local geographic, demographic, historical, and economic context of places within Blue Economy conversations.

Although we hypothesised that there would be relatively low social carrying capacities in all three regions due to (a) the predominance of finfish aquaculture in Downeast and (b) the more gentrified (wealthy) communities in Damariscotta and Blue Hill, our results indicate a moderately high level of acceptance of marine aquaculture across all regions, albeit with a relatively lower support and more conflicts in Blue Hill compared to the other two regions (Figure 5.2, Tables 5.2–5.3).

From our analysis, we identify at least three different value frames associated with marine aquaculture: (1) Working Waterfront and Artisanal Compatibility, seen in Damariscotta; (2) Aesthetic Incompatibility, seen in Blue Hill; and (3) Workforce and Economic Development Compatibility, seen in Downeast. These differences are shaped by the regional demographic, economic, and historical context, as well as the values and aesthetic preferences of the three communities.

The high level of support for aquaculture observed in the Damariscotta, a relatively wealthy and gentrified region (Table 5.1), appears to be due to local perceptions of shellfish aquaculture as compatible with the region's small-scale, working waterfront and cultural heritage. In this region, coastal residents view shellfish aquaculture as fitting into their aesthetic and recreational ideals. Shellfish farming fits an image of small-scale, artisanal, local food production, and coastal residents in Damariscotta view it in much the same way they see lobstering, which is common throughout coastal Maine (Hanes, 2018). The quaint, working waterfront culture dominated by lobstering is part of the landscape that drew amenity migrants to Maine's coast (Thompson et al., 2016). The Damariscotta case shows that gentrified regions can be supportive of aquaculture under the right circumstances.

In contrast, the relatively lower social carrying capacity of aquaculture in Blue Hill appears linked to differences in the way coastal residents view aquaculture. Demographically, the region is much like the Damariscotta: it has high levels of gentrification and extensive shellfish aquaculture (Table 5.1). Instead, aquaculture is not viewed as compatible with the landscape that brought these amenity migrants to Blue Hill; i.e. a working waterfront is not a priority for them. Much of the opposition in Blue Hill derives from residents seeking the national park's picturesque beauty and public hearing testimony in this region illustrates the preference for keeping local waterways in a pristine state (Hanes, 2018). Threats to the region's values and landscape preferences, initially with a salmon farm proposal, galvanised opposition that continued against subsequent oyster farms proposed. Unlike in Damariscotta, coastal residents in this region appear unable to fit shellfish aquaculture into their recreational and wilderness-based aesthetic preferences. This is consistent with others who have found that the ability of natural resource sectors to fit into gentrifiers' value frame or idealisations is a key determinant of their ability to expand production in gentrifying regions (Abrams and Bliss, 2013). The contrast between Downeast and Damariscotta illustrates that gentrifiers are not a homogenous group; clearly, the amenity migrants in

Blue Hill have different values related to aquaculture than those in Damariscotta. Blue Hill shows that small groups of wealthier individuals or those with high social capital can influence the social carrying capacity of aquaculture in a region.

Support for aquaculture in the Downeast region was especially unexpected given that the predominant species grown is Atlantic salmon, which tends to be perceived as having higher environmental impacts compared to other species (Johnson and Hanes, 2018). In this region, support for industrial aquaculture is influenced by the broader socioeconomic and historical context of this region; one characterised as significantly rural and continues to experience effects from an ongoing post-industrial transition. While a single foreign company owns all of the farms, the company does employ residents and is viewed locally as having supported economic development and jobs. The history and legacy of the region as a centre of the US sardine canning industry is also notable; an industrial workforce did not dominate the other two regions' history. Finfish aquaculture fits in with the region's workforce and economic development preferences, and so it makes sense that the region would like to see more growth in this sector (Figure 5.2). A compatible workforce and economic development frame also drive support in this region for other industrial sectors, like marine renewable energy (Johnson et al., 2013). Unlike Damariscotta, which values and embraces the small-scale, farmer-friendly aquaculture farmer as part of its working waterfront, the Downeast region also values the large-scale, industrial aquaculture model that salmon farming contributes to the working waterfront.

The historical development of aquaculture is also central in each of the regions. It is not insignificant that within both Damariscotta and Downeast regions are Sea Grant and Cooperative Extension staff who work closely as boundary spanners (Cash and Moser, 2000) between the community and the industry, while also helping the industry via technology transfer. It is also important that the development in Damariscotta and Downeast started with a few small growers and then expanded in a way that was integrated into the community. Downeast eventually consolidated, but the early history was one of small-scale, family farmers. In contrast, serious discussion about aquaculture in Blue Hill started with a very large proposal for a salmon farm that did not originate within the community. The path dependencies from these early histories are significant in shaping acceptance levels today. These experiences suggest that where aquaculture is newly emerging, efforts are needed to engage the public in planning discussions before proposals get "too big, too quickly" and such planning and discussions can benefit from boundary spanners like those from Sea Grant and Cooperative Extension.

The contrast between these regions shows that demographic explanations for social carrying capacity are insufficient and neither surveys nor public hearings alone provide the necessary context for understanding conflicts and acceptance of marine aquaculture. Instead, we argue for combining multiple types of marine social science as there are no single indicators that effectively predict conflicts. Socioeconomic data can provide context needed to contextualise conflicts in a given region, but differences in communities' values' perceptions are hard to generalise based on such data. This is seen most clearly with the contrast between Damariscotta and Blue Hill. Survey data are relatively easy to collect and are able to capture and quantify broad differences in values and opinions among residents across regions, while also giving representation to those not able or interested to participate in public governance processes. Understanding localised conflict driven by vocal subsets of the community, such as the amenity migrants in Blue Hill, can be clearly captured in an analysis of lease hearings transcripts. Public hearing testimony is a common data source, given the extent of participatory processes in developed countries and as such, we view it as a valuable and underused source for understanding drivers of social carrying capacity. Finally, ethnographic research was critical for revealing complex nuances in the values held by residents and the patterns driving conflicts that were not visible in the quantitative hearing analysis or survey data. As well, qualitative research was necessary for the design of the structured mail survey. It is the combination of marine social science data that allow us to best understand the place-based, social carrying capacities of marine aquaculture. In this way, this study illustrates how "creative combinations of theories and methods" (Krause et al., 2015) reveal complex and nuanced aspects of marine aquaculture, and revealing further the opportunity and need to integrate marine social science into Blue Economy agendas (see McKinley's Chapter 2).

To meet its global potential as part of the Blue Economy, marine aquaculture production needs to expand to new regions. Conflicts are likely to emerge in some, but not all areas. Predicting *a priori* which areas are going to experience conflict based on demographic and socioeconomic indicators alone is insufficient. Instead, marine social science research is needed to understand the values and other conditions setting the limits for marine aquaculture in order to advance this increasingly critical component of the Blue Economy.

Acknowledgements or Credits List

This project was supported by the USDA National Institute of Food and Agriculture, Hatch project number #ME022012 through the Maine Agricultural & Forest Experiment Station and the National Science Foundation award #IIA-1355457 to Maine EPSCoR at the University of Maine.

Note

1 Data are available at https://www.maine.gov/dmr/aquaculture/maine-aquaculture-leases-and-lpas/maine-aquaculture-leases-and-lpas.

References

Abrams, J., & Bliss, J. C. (2013). Amenity landownership, land use change, and the re-creation of "working landscapes". *Society & Natural Resources*, 26(7), 845–859.

Bricknell, I. R., Birkel, S. D., Brawley, S. H., Van Kirk, T., Hamlin, H. J., Capistrant-Fossa, K., ... & Moeykens, S. (2021). Resilience of cold water aquaculture: a review of likely scenarios as climate changes in the Gulf of Maine. *Reviews in Aquaculture*, 13(1), 460–503.

Cash, D. W., & Moser, S. C. (2000). Linking global and local scales: designing dynamic assessment and management processes. *Global Environmental Change*, 10(2), 109–120.

Cleaver, C., Johnson, T. R., Hanes, S. P., & Pianka, K. (2018). From fishers to farmers: assessing aquaculture adoption in a training program for commercial fishers. *Bulletin of Marine Science*, 94(3), 1215–1222.

Creswell, J. W., & Clark, V. L. P. (2014). *Designing and Conducting Mixed Methods Research.* Sage Publications: Thousand Oaks, CA.

Dalton, T., Jin, D., Thompson, R., & Katzanek, A. (2017). Using normative evaluations to plan for and manage shellfish aquaculture development in Rhode Island coastal waters. *Marine Policy*, 83, 194–203.

Dillman, D. A., Smyth, J. D., & Christian, L. (2014). *Internet, Phone, Mail, and Mixed-Mode Surveys: The Tailored Design Method.* John Wiley & Sons: Hoboken, NJ.

Gentry, R. R., Froehlich, H. E., Grimm, D., Kareiva, P., Parke, M., Rust, M., Gaines, S. D., & Halpern, B. S. (2017). Mapping the global potential for marine aquaculture. *Nature, Ecology, & Evolution, 1*, 1317–1324.

Gibbs, M. T. (2009). Implementation barriers to establishing a sustainable coastal aquaculture sector. *Marine Policy*, 33, 83–89.

Hanes, S. P. (2018). Aquaculture and the postproductive transition on the Maine Coast. *Geographical Review*, 108, 185–202.

Johnson, T. R. (2020). Reflecting on Maine's changing productive coastal region. *Maine Policy Review*, 29(2), 91–97.

Johnson, T. R., & Hanes, S. P. (2018). Considering social carrying capacity in the context of sustainable ecological aquaculture, in Patrick Heidkamp, C. & Morrissey, J. (Eds.) *Towards Coastal Resilience and Sustainability* (171–187). Routledge: Oxon.

Johnson, T. R., Beard, K., Brady, D. C., Byron, C. J., Cleaver, C., Duffy, K., Keeney, N., Kimble, M., Miller, M., Moeykens, S., Teisl, M., van Walsum, G. P., & Yuan, J. (2019). A social-ecological system framework for marine aquaculture research. *Sustainability*, 11, 1–22.

Johnson, T. R., Jansujwicz, J. S., & Zydlewski, G. (2013). Tidal power development in Maine: stakeholder identification and perceptions of engagement. *Estuaries and Coasts*, 38, 266–278.

Kluger, L. C., Filgueira, R., & Byron, C. J. (2019). Using media analysis to scope priorities in social carrying capacity assessments: a global perspective. *Marine Policy*, 99, 252–261.

Knapp, G., & Rubino, M. C. (2016). The political economics of marine aquaculture in the United States. *Reviews in Fisheries Science and Aquaculture*, 24, 213–229.

Krause, G., Brugere, C., Diedrich, A., Ebeling, M. W., Ferse, S. C. A., Mikkelsen, E., Pérez Agúndez, J. A., Stead, S. M., Stybel, N., & Troell, M. (2015). A revolution without people? Closing the people-policy gap in aquaculture development. *Aquaculture*, 447, 44–55.

Maine Department of Marine Resources. 2018. Aquaculture in Maine. Available at https://www.maine.gov/dmr/aquaculture

McKindsey, C. W., Thetmeyer, H., Landry, T., & Silvert, W. (2006). Review of recent carrying capacity models for bivalve culture and recommendations for research and management. *Aquaculture*, 261, 451–462.

Thompson, C., Johnson, T., & Hanes, S. (2016). Vulnerability of fishing communities undergoing gentrification. *Journal of Rural Studies*, 45, 165–174.

U.S. Census Bureau (2017). *American Community Survey 5-year estimates.* Available at https://data.census.gov/cedsci/

Walters, B. B. (2007). Competing use of marine space in a modernizing fishery: salmon farming meets lobster fishing on the Bay of Fundy. *The Canadian Geographer/Le Géographe canadien*, 51, 139–159.

Weitzman, J., & Filgueira, R. (2020). The evolution and application of carrying capacity in aquaculture: towards a research agenda. *Reviews in Aquaculture, 12*(3), 1297–1322.

Woodward, C. (2015). *The Lobster Coast: Rebels, Rusticators, and the Struggle for a Forgotten Frontier.* New York: Penguin.

6

OF FRAGILE COMMUNITIES AND BIG DREAMS

The Finnafjörðurur Harbour Project in Northeast Iceland

Matthias Kokorsch and Johannes Stein

6.1 Introduction

As the Arctic continues to warm and the sea ice continues to recede, arctic shipping is moving into the spotlight. Shortened routes between East Asia, the East coast of North America, and Europe promise to reduce transport time, emissions, and fuel costs (Smith & Stephenson, 2013; Melia et al., 2016). However, most transportation through the Arctic requires ice-proven vessels that are less fuel-efficient than their conventional counterparts. Based on current developments, the Finnafjörður port-project was born. At the edge of the Arctic in Northeastern Iceland, a transit harbour is to be built. In a hub-and-spoke system, ice-proven vessels traversing the Arctic can transfer their cargo to lighter ships that then finish the journey to the markets in Europe and North America (Hagfræðistofnun Háskóli Íslands, 2019).

The idea for the Finnafjörður port-project dates back to 2007 when the two municipalities in the area, Langanesbyggð and Vopnafjarðarhreppur, requested the Icelandic engineering company EFLA to conduct preliminary assessments of the envisaged location. In 2012 the German harbour development company Bremenports joined the project, and a contract was signed between the four parties in 2014. While the project is first and foremost a harbour construction project, it also intends to improve infrastructure in the region (roads, electricity, etc.) and establish a tax- and duty-free zone (Bremenports, 2015). Construction was supposed to start in 2021 but has been postponed to 2025. As part of the project the Finnafjörður port development company was founded, which oversees the development of the port and manages the harbour. Bremenports holds the majority of shares while smaller amounts are allocated to EFLA and the municipalities. The two municipalities own the Finnafjörður port authority, share responsibilities, and will operate the harbour once the construction is completed.

DOI: 10.4324/9781003280248-8

As part of the development process, it is intended that concessions are made to third parties. These concessionaires will develop parts of the project and will get certain rights for future usage (Bremenports 2015, 2019).

Considering the growing attention for new shipping routes in the Arctic (see Panahi et al., 2021), it is surprising that the Finnafjörður project has somewhat been sidelined, both in domestic coverage and international publications. This is remarkable given the potential impact of one of the largest infrastructure projects in Icelandic history. Port-projects in remote and rural settings are rarely addressed in port research and a critical/cultural geography approach is even harder to find (Ng et al., 2014, Panahi et al., 2021). In addition, local voices are usually not given much space in discussions on Arctic infrastructure projects (Pahl & Kaiser, 2018).

The Finnafjörður port-project changes the paradigm of scale: While the Icelandic utilisation of the Ocean has been connected to extractive fisheries for the past 120 years in its surrounding waters, such a project shifts the focus to new scales and global structures. The region around Finnafjörður would become a new geographical focal point in international trade, instead of being a captive of "double-peripheralisation": one of the remotest regions on a remote island. The question is: to what extent is such a paradigmatic and scale-related shift desirable? This chapter addresses this question and asks further: whether such a port-project increases community resilience and fosters a just-transition for struggling and mono-industrial communities? We thus focus on local perceptions and knowledge as well as potential coastal community impacts, as suggested by McKinley et al. (2020). We will also answer the question, whom this project is potentially for, and whom not (McKinley, Chapter 2, this volume). Potential conflicts will be identified between different groups within the three villages.

For answering parts of these questions, a survey was conducted in the villages. The theoretical foundations for the interpretation of the survey results are community resilience and the Blue Economy. Attention is put on aspects of transformation, reorientation, adjustment, exogenous, and endogenous growth.

6.2 Blue Economy and Community Resilience

The main theoretical concepts for this chapter are community resilience and Blue Economy. Before the two main concepts for this case study—community resilience and the Blue Economy—will be further defined, a brief overview over the past 40 years of regional and coastal community development in Iceland is necessary for contextualisation.

Iceland has been an important fishing nation since the early 1900s and most boom-and-bust cycles of coastal villages, especially since the mid-1930s and the industrialisation and advancing technology of the fishing industry, can be linked directly to this industry (Kokorsch & Benediktsson, 2018a). The threat of overfishing and the possible loss of this precious resource eventually led to the introduction of an ITQ system (individual transferable quotas). The de-facto privatisation of the fisheries changed the economic structure of many fishing-dependent villages rapidly. Quota concentration and consolidation took place (Agnarsson et al., 2016) and came with undesirable social side effects (Kokorsch & Benediktsson, 2018b). Having lost the mainstay industry, many coastal villages had to readjust their economic structure. In some villages the strategy was to keep fisheries at all costs, others diversified within the fisheries sector, while others turned to completely unrelated industries. One of the main strategies for large-scale regional development was the building of heavy industries, such as aluminium smelters, exploiting the other natural resource Iceland has on offer: renewable energy (Benediktsson, 2014).

While the success of the quota system, diversification strategies, and large-scale development projects during the past 40 years can be debated, one major challenge still exists nonetheless: urbanisation. The main reasons for adolescents to leave are educational and lack of job-opportunities (Bjarnason, 2014). Many small and remote places in Iceland have serious demographic problems (Kokorsch & Benediktsson, 2018a) and the gender balance in many villages is skewed.

For the setting of this study—remote coastal villages in Iceland—the following definition of community resilience is suitable:

> Community resilience is the ability of a community to cope and adjust to stresses caused by social, political, and environmental change and to engage community resources to overcome adversity and take advantage of opportunities in response to change. (Amundsen, 2012, p. 1)

The individual components of the definition—social, political, and environmental changes or stressors—can be identified in the recent history of Icelandic regional and coastal community development. In addition, the search for endogenous resources, opportunities, and pathways to overcome adverse socio-economic and demographic developments is also prominent in Iceland. Earlier research has already explored the concept of community resilience in previous case studies in Iceland (Kokorsch & Benediktsson, 2018b)—these case studies have provided insight into the assessment of community resilience and provided seven components which have been applied to the Finnafjörður port-project and the affected communities (see Table 6.1).

For connecting community resilience and the Blue Economy, we follow the conceptualisation of Garland et al. (2019) and their elaborations on geographical concepts for framing a sustainable Blue Economy. The main approaches for our interpretation are Space and Place, Proximity and Distance, and Social Carrying Capacity. The latter can be related to the question of potential winners and losers of transformation and coping strategies, in this case large-scale development projects and a major reorientation. Garland et al. (2019, p. 14) state: "Resulting conflicts about the social acceptance of certain aspects of development must not be understood as homogenous across space, but instead as geographical processes that change through different understandings of economy, local distinctiveness,

TABLE 6.1 Resilience and transformation components derived from case studies in Iceland (table modified from Kokorsch & Benediktsson, 2018b, p. 100)

Component	*Description/Definition*
Coping strategy	Innovation, networking, and formation of identity; new solutions to local problems
Local agency	Destiny in own hand; ability to shape own future
Place attachment/Sense of belonging	Emotional ties; being open and welcoming; encouraging participation in community
Transformation	Ability to create new pathways and enhance functioning
Community infrastructure	Services provided, to people and industries
Driver of change	Public or private
Diversified economy	No over-reliance on a single industry

attachments to places, and ideas of social carrying capacity." Regarding proximity and distance, the port-project affects all four dimensions of distance: spatial, temporal, social, and hypothetical distance (cf. Garland et al., 2019). Future decisions might no longer be made within the vicinity of the villages and region; hence, local agency might decrease. Remote settings at the edge of the Arctic and the possibility for new trading routes and resource exploitation may eventually lead to some sort of time-space compression and the "erosion of local distinctiveness" (Harvey, 1989; Garland et al., 2019). Altering the local distinctiveness may affect the resilience-building components place attachment and sense of belonging substantially.

6.3 Fragile Communities: A Case Study

6.3.1 The Case Study Location and the Project "Fragile Communities"

Against the background of community resilience and the Blue Economy, one community development strategy in Iceland needs to be introduced. In 2012, the Icelandic Regional Development Institute launched the project "Fragile Communities" (icel.: Brothættar byggðir). This project is an important paradigmatic shift, since regional development in Iceland was often connected to large-scale industrial development and top-down approaches (cf. Seyfrit et al., 2010; Kokorsch & Benediktsson, 2018b). This project is about promoting endogenous strategies and civic engagement, supporting small-scale projects from within the community, and eventually fostering the collective/self-efficacy of the local population.

The communities apply for this scheme themselves and will then get up to five years of governmental support through the Regional Development Institute. Until 2020, 12 communities have been participating, all but one being former fishing villages, that had problems to readjust after the loss of fishing rights. The success of the project varies between communities and the definition of success.

TABLE 6.2 Population development, female population (in %), and average age in 1991 and 2019 in the three villages. Source: Own calculations based on Statistics Iceland (2021)

	Inhabitants		*Average age*		*Female population*	
	1991	*2019*	*1991*	*2019*	*1991*	*2019*
Iceland	259,581	356,991	32	37	49.8	48.8
Þorshöfn	416	369	36	43	45.9	43.6
Bakkafjörður	99	69	57	71	41.4	47.8
Vopnafjörður	689	527	36	45	48.2	51.0

One of the communities is Bakkafjörður, which is located close to the envisaged harbour location of the Finnafjörður port-project.

Finnafjörður itself is located in the Northeast of Iceland (see figure 6.1). The port-project is located between two communities, Bakkafjörður and Þorshöfn. A third one, Vopnafjörður, is involved due to its vicinity and administration. The demographic profile of the three communities can be seen in Table 6.2.

All three communities had a monotonous economic structure and have been dependent on fisheries. Since 1990—and the introduction of ITQ—the three communities had to readjust to a certain extent. Regarding demographic key figures, all three villages face the same challenges: outmigration, ageing above the national average, and a skewed gender balance.

Bakkafjörður has experienced the most negative population trend and is part of the fragile community project since 2019. The main aims are an attractive image of the community and a lively society in combination with the support of locally driven business ideas (Byggðastofnun, 2021).

6.3.2 The Finnafjörður Port-Project and Current Developments

Arctic vessel traffic is growing constantly and predicted to increase (Stevenson et al., 2019). The Finnafjörður port-project is based on scenarios regarding Arctic shipping opportunities in the coming decades. The scenarios, namely, the Trans-Polar-Route (TPR) and the Northern Sea Route (NSR),[1] come with some uncertainties and obstacles. It can be debated in general, to what extent harbours increase local economies and employment (Ferrari et al., 2012; Elburz et al., 2017). Indirect costs and externalities need to be addressed in such a discussion (Arbués et al., 2015; Efimova and Gapochka, 2020).

A frequent debate is about the necessity of such a port and the future of Arctic shipping. While there are numerous voices proclaiming the growth potential for Arctic shipping and related companies, there are also more and more cautious voices. One of the main reasons for questioning the port-project is summarised by Bennett et al. (2020, p. 4), who state that "[...] much of the recent growth in Arctic shipping has been destinational, involving the transportation of cargo to Arctic locations and of resources out of the region, rather than transit, or using

Arctic waterways to move cargo between two non-Arctic ports." A harbour for such activities is not needed in Finnafjörður and recent estimations for destinational shipping through the TPR are low (Stevenson et al., 2019; Bennett et al., 2020). The only economically viable option for Finnafjörður would be a hub-and-spoke system in combination with (offshore) resource extraction in form of oil and gas as well as shipping industries (Bennett et al., 2020). However, Finnafjörður has strong competitors with Longyearbyen for the TPR (Bennett et al., 2020) and Russia that heavily invests in ports along the NSR (Panahi et al., 2021).

In case the TPR or NSR become feasible and shipping activities continue to grow, the provision of infrastructure is important, especially for the TPR—and a lesser extent also the NSR (Wang et al., 2019). There are good reasons for locating such a harbour in Finnafjörður, building it up from scratch, given the poor infrastructure, status, and development potential of comparable settings and existing harbours (cf. Wang et al., 2019).

As mentioned in previous sections, Iceland and its transportation and economy were heavily linked to fisheries for most of the 20th century and were otherwise too far off in geographical sense. In addition, Iceland has a limited domestic market: scale and location were a hindrance in almost every respect. With the current developments, it could profit from advances in transportation, communication, technology, and international trade (cf. Ng et al., 2014). Regarding the TPR, Finnafjörður could be one of the first, largest, and thus most important harbours in the long run, incorporating three phases of port development—setting, expansion, specialisation—at a stroke, which comes with several advantages (cf. Monios & Wilmsmeier, 2012). The region around Finnafjörður could possibly profit from future developments in the Arctic and getting a share of a process that will happen regardless a port being built in Iceland or elsewhere (Pahl & Kaiser, 2018). The region could thus benefit from comparative and competitive advantage (Bottasso et al., 2014).

A recent report on transportation issues indicates that transportation in Icelandic is either tied to cruise ship tourism or inland travel, but not at all to cargo shipping and new transportation routes. This is not surprising considering the fact that ports need markets, people, and cargo flows (Efimova and Gapochka, 2020). All three variables are missing in this sparsely populated and remote region of Iceland. The frequently mentioned benefits for the hinterland of a port-region are another neglectable aspect for this location.

6.3.3 Methodology

Our research project had two steps. First, exploratory semi-structured interviews were conducted with key informants that are directly involved in the project as well as regional experts. Those deepened the overall insight of the researchers and were used in setting up the questionnaire. Interviews ranged from 60 to 90 minutes. Second, a questionnaire was created based on the research questions

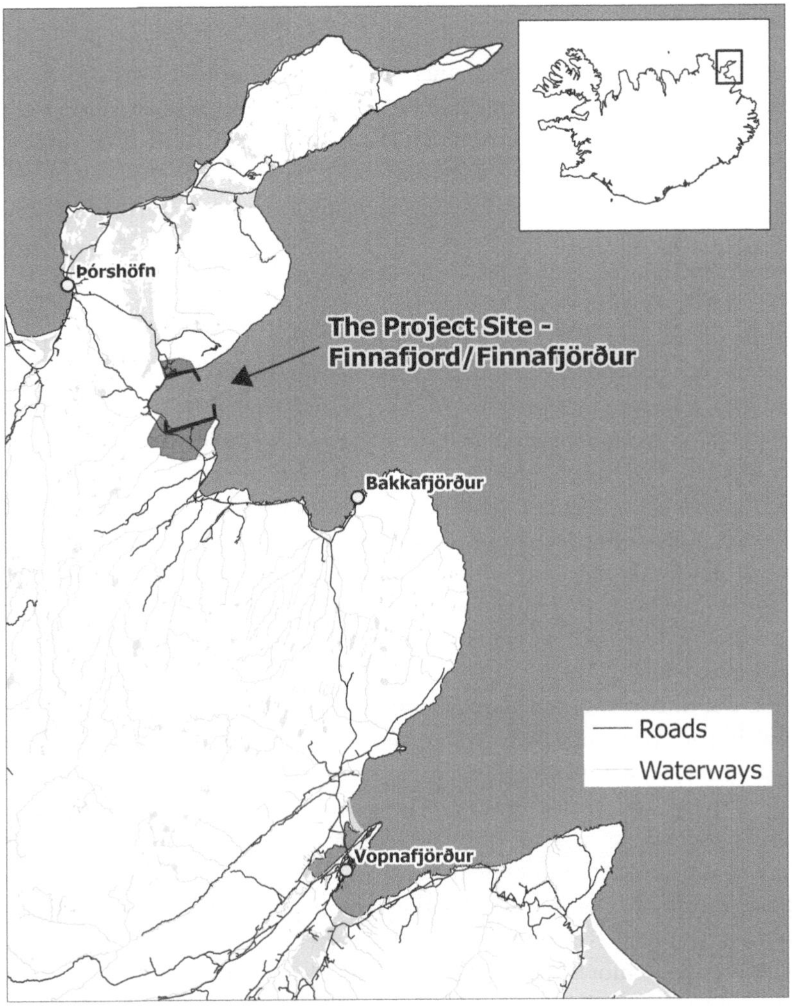

FIGURE 6.1 The Finnafjörður port-project site

and input from the interviews. In total, 23 open, multiple-choice, rate, and Likert scale questions about approval, impact, and agency of the project as well as socio-demographic basics were asked, following general survey guidelines (Dillman et al., 2014). The questionnaire was set up in English and Icelandic. Both versions were pre-tested by individuals from varying backgrounds for comprehensibility, visual appeal, and order before they were finalised. A paper and online version were set up. The researchers distributed the paper versions to all accessible households in the research area (see figure 6.1). The online version was advertised by repeatedly posting in local social media groups and utilising mailing lists from local contacts. Additionally, the researchers spent time in the field, promoting the survey in person. The survey ended after four weeks.

6.3.4 Findings

Overall, 96 surveys from inhabitants in the region around Finnafjörður could be used for further analysis. In general, the port-project is polarising, with most respondents choosing the extreme ends of the approval scale. On a scale from 1 (full disapproval) to 10 (full approval of the project), 25% fully disapprove of the port-project while 30% fully approve it (mean value 5.7). An almost similar result can be found regarding the importance of the project for the region (mean value 6.0), with less respondents at the extreme poles of the scale.

The not-in-my-backyard phenomenon can be observed: The further away from the intended harbour location, the higher the overall approval found in the survey results. For the three communities, the approval is lowest in Bakkafjörður (mean 4.4), followed by Þórshöfn (mean 5.5) and Vopnafjörður (6.7). Significant discrepancies can be revealed when demographic variables are considered, especially gender. The mean value among female respondents is 4.7 and 6.8 among male respondents. A closer look into the reasons for or against approval delivers a possible explanation for this striking discrepancy. The main reasons for disapproval among female respondents are environmental concerns and the loss of regional identity. In return, possible economic activities in the region and infrastructure development were the most prominent reasons for approval among male respondents.

Respondents were asked about the magnitude (very weak to very strong) and overall impact (very negative to very positive) on a five-point scale regarding five components: environment, regional identity, population development, economic activity, and infrastructure). The average results for each variable can be seen in Table 6.3.

In another question, respondents were asked whether the port-project is overall positive for the region: 20% of the respondents disagreed and 44% agreed with this statement. Regarding the positive effects on the individual livelihood, respondents were more pessimistic with 28% disagreeing and 22% agreeing.

One of the most pressing issues in the region is out-migration and an undiversified labour market from the coastal communities. Respondents were asked to consider whether the port-project delivers a new solution to this challenge.

TABLE 6.3 Mean value for the five aspects to be graded after the strength of the impact (from 0 "no impact at all" to 4 "very strong impact") and the nature of the impact, ranging from −2 (very negative) to 2 very positive)

	Strength of impact (0:4)	*Nature of impact (−2:2)*
Environment	2.8	−0.7
Population development	3.1	1.0
Economic activity	3.0	1.2
Infrastructure	3.2	1.0
Regional identity	3.0	−0.2

Analysis found that 43% of the respondents do not consider a job in the new harbour in the future, compared with 17% of respondents who indicated that it might be an opportunity. The results look completely different when asked whether they feel jobs might be attractive for other inhabitants within the region: only 11% (strongly) disagree with that statement compared with 53% that (strongly) agree.

An interesting observation is that the origin of the respondents mattered regarding the overall approval of the port-project. Analysis found that it is mainly those growing up elsewhere in Iceland being in favour of the port-project (mean 7, on a scale from 1 to 10) while those who grew up in the region of Finnafjörður show significantly lower approval (mean 5.1).

Aspects of scale matter for the respondents and are polarising: 34% of the respondents (strongly) disagree with the statement that the port-project is too big for the region, while 32% (strongly) agree with the statement. However, the port-project is seen rather as an addition and improvement than a threat to current infrastructure. The expected magnitude of the impact is highest for the infrastructure and 52% expect (very) positive impacts (see Table 6.3). Considering the occupational background, no objections from respondents from the local fish-industry can be revealed.

Respondents were asked about the influence of stakeholders in this private-public-partnership project and to what extent the project partners have an impact on the overall approval. The results show that the project-partners are comparatively irrelevant for the overall approval of the project. However, the influence of the private partners is considered as too much. The private partners Bremenports and EFLA are considered as (too) influential (53% and 40%, respectively). Apparently, it makes a difference for the acceptance, whether it is an Icelandic or international project partner. In addition, some 48% would like to be more involved as individuals.

6.4 Conclusion

Having identified components for the identification and evaluation of resilience-building strategies in coastal communities of Iceland, we will now address those and interpret them against the background of the survey results (see Table 6.4). The port-project is very polarising and the question as to whether it could be a possible coping strategy cannot be answered. On the one hand, the port-project comes with some scale-related challenges and is not necessarily a new solution to old problems in the region. It rather reflects an early and thus mainly economically driven approach within the evolutionary journey of the Blue Economy as described by McKinley (Chapter 2, this volume). A diversified economy and successful transformation might result in positive outcomes as expected by respondents; however, these may also result in high environmental impacts and do not seem to be attractive for female inhabitants. That economic benefits and enhanced functioning are considerably lower than previously expected has been

TABLE 6.4 Resilience and transformation components applied to the Finnafjörður port-project (table modified from Kokorsch & Benediktsson, 2018b, p. 100)

Component	*Description/Definition*	*Applied to Finnafjörður*
Coping strategy	Innovation, networking, and formation of identity; new solutions to local problems	New solutions or new challenges?
Local agency	Destiny in own hand; ability to shape own future	New dependencies or self-determination?
Place attachment/Sense of belonging	Emotional ties; being open and welcoming; encouraging participation in community	What happens to the social fabric and place identity? Loss of local distinctiveness?
Transformation	Ability to create new pathways and enhance functioning	What sort of pathways are envisaged? Embeddedness?
Community infrastructure	Services provided, to people and industries	Chances or displacement?
Driver of change	Public or private	New dependencies through exogeneous strategies?
Diversified economy	No over-reliance on a single industry	New dependencies through mono-structural economy

observed in comparable settings and projects within Iceland. Conversely, environmental impacts are usually stronger than expected during the planning phase. However, it does not seem to impact the overall approval.

Regarding the resilience-building components' place attachment and sense of belonging, we conclude that the changes of the place identity and local distinctiveness are arguments against the port-project. That almost 40% of the respondents expect (very) negative consequences for the regional identity should not be underestimated. An interesting result can be connected to the components' local agency and driver of change. The threat of new dependencies—from an agency perspective and regarding external stakeholders—cannot be identified in this survey. For the respondents, it does not seem to make a significant difference regarding the approval, whether government agencies or the private industry are the main drivers of change. So far, the focus of comparable projects has been on the local communities as objects, somewhat sidelining aspects of self-determination as well as local needs (Pahl & Kaiser, 2018). One can argue that the question who participates in decision-making and who benefits from the economic wealth generated is a central element in the discussion (Bennett et al., 2019). For Finnafjörður, a public-private partnership is envisaged, which seems to be common for infrastructure investments in the Arctic (Bennett et al., 2020). We expected more potential conflicts, since "[…] changing ownership structures in port operations, especially the entrance of private equity firms, raises the very real danger of 'resource stripping' rather than investment" (Ng et al., 2014, p. 7). Yet, the question whether profits are privatised, and losses socialised, seems not to be of importance at this point.

This is even more surprising when the distribution of costs is considered. It can be expected that the main costs associated with the port-project—directly or indirectly—are covered by the municipalities and the Icelandic state, which usually carries some potential for conflicts (Ferrari et al., 2012; Bottasso et al., 2014). Especially when scarce economic resources are tied to infrastructure and development, more endogenous and small-scale strategies might lack funding opportunities. Until 2021 at least 450 million ISK (approx. 3.5 million US$) have been invested in this project by the Icelandic state and the companies involved (Bennett et al., 2020). In addition, another 18 million ISK (approx. 140,000 US$) have been added from the Icelandic Regional Development Agency for continuing the preparatory work. One may ask whether these investments could not have been targeted at real endogenous strategies, meeting local demands, encouraging a just-transition of a truly inclusive character—like the project Fragile Communities suggests. It is thus not surprising to see the lowest rate of approval in Bakkafjörður. The port-project does not seem to be a promising coping strategy when those groups needed the most in a depopulating region are considered: women and young people.[2] The gender aspect should not be underestimated in this regard and the results from the survey show the importance to address this issue further. It seems that women neither have the interest to work in a hitherto male-dominated fish industry nor in potential large-scale industrial complex. The port-project might lead to a sudden influx of people, but it is most likely not a solution for balancing a skewed gender distribution in the region.

Considering the obstacles and challenges of the port-project, the possible returns for the region are rather vague. The port-project might not be the best tool for a just-transition and the question of social carrying capacity remains a pressing issue: the port-project could overstretch the carrying capacity. A place that certainly needs economic and demographic stimuli might end up overstimulated, compensating the loss of a few, with the fluctuation of many. Three once lively fishing villages could be turned into a placeless harbour, losing its distinctiveness (cf. Garland et al., 2019). This relates to another aspect for resilience building, transformation, and reorientation: Embeddedness, or the question to what extent such a project fits the local socio-cultural environment (Ng et al., 2014). From the responses in the survey, one can conclude that the port-project does not fit the grown structures of the region and the attached industries. Against this background, one can understand why this port-project is hardly addressed in the first report regarding the fragile community project in Bakkafjörður. In fact, these are merely ecological concerns that are brought up by the inhabitants when talking about Finnafjörður (Byggðastofnun, 2021).

All in all, at this stage, the port-project does not recognise the "specific needs, priorities and challenges of environments, people and places at a range of scales" (McKinley, Chapter 2, this volume, pp?). The development in comparable settings in other Icelandic communities might serve as a warning example. This leads into the interregional aspect of a just-transition: the coastline in the Northeast and East of Iceland already has some harbours, of which two were tailored solely to heavy industries. From experience, it can be doubted that all harbours will be economically viable in the long run (cf. Monios & Wilmsmeier, 2012).

Notes

1 For an extensive summary of potential options and scenarios, including the Northwest Passage, the Northern Sea Route, and the TPR, see Pahl and Kaiser (2018).
2 For comparison, the fragile community project in Bakkafjörður received 16 million ISK for funding endogenous project ideas.

References

Agnarsson, S., Matthiasson, T., & Giry, F. (2016). Consolidation and distribution of quota holdings in the Icelandic fisheries. Marine Policy, 72, 263–270.

Amundsen, H. (2012). Illusions of resilience? An analysis of community responses to change in northern Norway. Ecology and Society, 17(4), 46.

Arbués, P., Baños, J. F., & Mayor, M. (2015). The spatial productivity of transportation infrastructure. Transportation Research Part A: Policy and Practice, 75, 166–177.

Benediktsson, K. (2014). Nature in the 'neoliberal laboratory'. Dialogues in Human Geography, 4(2), 141–146.

Bennett, N. J., Cisneros-Montemayor, A. M., Blythe, J., Silver, J. J., Singh, G., Andrews, N., … & Sumaila, U. R. (2019). Towards a sustainable and equitable Blue Economy. Nature Sustainability, 2(11), 991–993.

Bennett, M. M., Stephenson, S. R., Yang, K., Bravo, M. T., & De Jonghe, B. (2020). The opening of the Transpolar Sea Route: Logistical, geopolitical, environmental, and socioeconomic impacts. Marine Policy, 121, 104178.

Bjarnason, T. (2014). Adolescent migration intentions and population change: A 20-year follow-up of Icelandic communities. Sociologia Ruralis, 54(4), 500–515.

Bottasso, A., Conti, M., Ferrari, C., & Tei, A. (2014). Ports and regional development: A spatial analysis on a panel of European regions. Transportation Research Part A: Policy and Practice, 65, 44–55.

Bremenports. (2015). Finnafjörður Port—Project. Retrieved 28 May 2020, from https://bremenports.de/Finnafjörður

Bremenports. (2019). Hafen für nachhaltige Entwicklung im Nordosten Islands. Retrieved on 2 June 2020, from https://bremenports.de/hafen-fuer-nachhaltige-entwicklung-im-nordosten-island/

Byggðastofnun. (2021). Markmið og framtíðarsýn, samþykkt á íbúafundi í nóvember 2019. Retrieved from: https://www.byggdastofnun.is/static/files/Brothaettar_byggdir/Bakkafjordur/betri-bakkafjordur-lokaskyrsla-12.-november-2019.pdf

Dillman, D. A., Smyth, J. D., & Christian, L. M. (2014). Internet, phone, mail, and mixed-mode surveys: The tailored design method. John Wiley & Sons: Hoboken, NJ.

Efimova, E. G., & Gapochka, A. A. (2020). Seaports as drivers of regional economic development: The case of Saint Petersburg and Leningrad Province. Case Studies on Transport Policy, 8(2), 383–393.

Elburz, Z., Nijkamp, P., & Pels, E. (2017). Public infrastructure and regional growth: Lessons from meta-analysis. Journal of Transport Geography, 58, 1–8.

Ferrari, C., Merk, O., Bottasso, A., Conti, M., & Tei, A. (2012). Ports and regional development: a European perspective. OECD Regional Development Working Papers, 2012/07, OECD Publishing.

Garland, M., Axon, S., Graziano, M., Morrissey, J., & Heidkamp, C. P. (2019). The Blue Economy: Identifying geographic concepts and sensitivities. Geography Compass, 13(7), e12445.

Hagfræðistofnun Háskóli Íslands. (2019). Skýrsla nr. C19:06 Siglingar á norðurslóðum—Ísland í brennidepli. Retrieved 5 June 2020, from http://hhi.hi.is/sites/hhi.hi.is/files/sjz/nordurslodir_lok_20_12_19_0.pdf

Harvey, D. (1989). The condition of postmodernity (Vol. 14). Blackwell: Oxford.

Kokorsch, M., & Benediktsson, K. (2018a). Prosper or perish? The development of Icelandic fishing villages after the privatisation of fishing rights. Maritime Studies, 17(1), 69–83.

Kokorsch, M., & Benediktsson, K. (2018b). Where have all the people gone? The limits of resilience in coastal communities. Norsk Geografisk Tidsskrift—Norwegian Journal of Geography, 72(2), 97–114.

McKinley, E., Acott, T., & Yates, K. L. (2020). Marine social sciences: Looking towards a sustainable future. Environmental Science & Policy, 108, 85–92.

Melia, N., Haines, K., & Hawkins, E. (2016). Sea ice decline and 21st century trans-Arctic shipping routes. Geophysical Research Letters, 43(18), 9720–9728.

Monios, J., & Wilmsmeier, G. (2012). Giving a direction to port regionalisation. Transportation Research Part A: Policy and Practice, 46(10), 1551–1561.

Ng, A. K., Ducruet, C., Jacobs, W., Monios, J., Notteboom, T., Rodrigue, J. P., ... & Wilmsmeier, G. (2014). Port geography at the crossroads with human geography: Between flows and spaces. Journal of Transport Geography, 41, 84–96.

Pahl, J., & Kaiser, B. A. (2018). Arctic port development. In Arctic marine resource governance and development (pp. 139–184). Springer: Cham.

Panahi, R., Ng, A. K., Afenyo, M., & Lau, Y. Y. (2021). Reflecting on forty years contextual evolution of arctic port research: The past and now. Transportation Research Part A: Policy and Practice, 144, 189–203.

Seyfrit, C. L., Bjarnason, T., & Olafsson, K. (2010). Migration intentions of rural youth in Iceland: Can a large-scale development project stem the tide of out-migration? Society and Natural Resources, 23(12), 1201–1215.

Smith, L. C., & Stephenson, S. R. (2013). New trans-Arctic shipping routes navigable by midcentury. Proceedings of the National Academy of Sciences, 110(13), E1191–E1195.

Statistics Iceland. (2021). Municipalities and urban nuclei. Retrieved from https://statice.is/statistics/population/inhabitants/municipalities-and-urban-nuclei/

Stevenson, T. C., Davies, J., Huntington, H. P., & Sheard, W. (2019). An examination of trans-Arctic vessel routing in the Central Arctic Ocean. Marine Policy, 100, 83–89.

Wang, D., Li, D., Gong, Y., Wang, R., Wang, J., & Huang, X. (2019). Development situation and future demand for the ports along the Northern Sea Route. Research in Transportation Business & Management, 33, 100465.

PART 2

Blue Economy—Regions

Blue Economies in Place

7

THE BLUE ECONOMY AND ITS GEOGRAPHIES

The Case of Turkey

Nuri Yavan and Mehmet Ragıp Kalelioğlu

7.1 Introduction

In recent years, the "Blue Economy"[1] concept has offered significant opportunities to environmentally preserve and exploit oceans, seas, lakes, and waters in the land and marine resources in accordance with sustainable development targets while providing economic growth at national and regional scales. In this context, both international organisations (EC, 2012, 2018, 2019, 2020; United Nations Conference on Sustainable Development, 2014; UNCTAD, 2014; UNEP, 2015; OECD, 2016; World Bank, 2017) and nation-states (e.g. for the USA, National Oceanic and Atmospheric Administration (NOAA), 2018; for Canada, Fisheries and Oceans Canada, 2020; for Ireland, Our Ocean Health, 2020; for China, Zhao et al., 2014) and researchers in the academic field (e.g. Morrissey and O'Donoghue, 2012; Colgan, 2013; Surís-Regueiro et al., 2013; Silver et al., 2015; Pinto et al., 2015; Doloreux, 2017; Winder and Le Heron, 2017; Voyer et al., 2018; Garland et al., 2019; Graziano et al., 2019; Heidkamp and Morrissey, 2019) have started doing a lot of researches on the "Blue Economy," which has become increasingly popular, both in the field of economic geography and other marine social sciences (McKinley et al., 2020).

A new sustainable development target has recently been determined with the pressure of developing technology and increasing competition conditions and the realisation of the significance of oceans, coasts, and inland waters as well as their proper use. In this context, on the one hand, the EU directs member countries to the Blue Economy in order to improve their current economy and support the green economy, on the other hand, it is seen that the UN directs the countries to the Blue Economy as a clean development model within the scope of sustainability of sea and ocean resources. Within this direction, environmental planning and governance of marine and coastal areas and that underdeveloped

DOI: 10.4324/9781003280248-10

and peripheral regions use of their potential and marine resources for local and regional development have provided the conversion of Blue Economy into a "strategic development tool", for both international organisations (EC, 2012, 2018; OECD, 2016) and some countries and local/regional authorities (Doloreux, 2017; Graziano et al., 2019). For that matter, the EU has defined Blue Economy as one of the top ten "emerging" industries in European countries since 2014 after it adopted the "Blue Growth Strategy" in 2012. Moreover, it has provided the identification of Blue Economy Clusters as "a priority domain" in more than 50 regions across the EU (Ketels and Protsiv, 2017) by making the concept part of smart specialisation strategies, the EU's regional development and innovation policy (De Vet et al., 2016; Doussineau et al., 2020).

Turkey provides a fairly interesting case for the analysis of the Blue Economy and its geographies. It is surrounded by sea on three sides, with a coastline of 8333 km (the fourth longest coastline in Europe after Norway, the UK, and Greece), and has a fairly big inland water and aquaculture composed of lakes, dams, rivers, and straits bridging two continents due to its geographical location, hundreds of ports (180), shipyards (78), a large fleet (15th big marine trade fleet in the World) (Karahan and Kirval, 2018), and rich aquatic and biological wealth. Turkey has great potential in terms of the Blue Economy and marine clusters and could be a sample country for analysis of the benefits and dynamics of blue growth. Accordingly, the primary goals of this study are both to support the studies in the agenda of Blue Economy determined by global institutions such as EU, OECD, World Bank, and UN and to contribute to the identification of sectors connected with the Blue Economy for a sea-based sustainable local and regional economic development in Turkey. To do so, this study will attempt to determine the sectors of the Blue Economy of Turkey and their place in the national economy focusing on Turkey, situated in the east of the Mediterranean and intense economic and social relationship with the EU. In other words, this study analyses the potential, performance, and size of Turkey's Blue Economy for the first time at a national scale using the classifications and definitions of the sectors under the Blue Economy framework of the EU and reveals its spatial distribution pattern across the regions of Turkey.

The study consists of five sections. After the introduction, the findings of the studies analysing the Blue Economy are presented to form the theoretical framework of the study. Following this is the third section which demonstrates data and method used in the study and how the analysis is conducted. This is followed by the findings section where the place of the Blue Economy in Turkey is demonstrated at a national scale and its geographical distribution is revealed through spatial analysis. In the last section, the findings are discussed in association with the related literature.

7.2 Blue Economy: Concepts and Frameworks

The concept of "Blue Economy" appeared as a consequence of the efforts of three different organisations (EU, UN, and NOAA). The EU, which used the concept in 2012[2] for the first time, accepted it as one of the most important tools of sustainable regional economic growth policy until the year 2020 by defining it as "the Blue Growth Strategy" (EC, 2012). In this respect, the EU has become a leader in the field by initiating the development of a variety of strategies for the transition to the Blue Economy, which requires a high level of innovativeness and technology in order to increase economic growth and employment. It has also steered academic studies with the reports it has prepared. In fact, the most important consequence of the EU's Blue Growth strategy is defining and categorising the industries forming the Blue Economy as of 2014 while generating data on blue industries and identifying "Blue Economy Clusters" in all EU regions based on classification and methodology developed by Ketels and Protsiv (2014, 2017) on behalf of EU, annual monitoring of development and growth dynamics of these blue clusters and supporting blue cluster initiatives. The report prepared by Ketels and Protsiv (2017) on behalf of the EU has been an especially significant step (Garland et al., 2019) and not only has it developed the EU's Blue Economy classification but it has featured the "clustering" dimension of the Blue Economy defining it in a spatial framework as well, and thus, enabled it to be defined as a primary field and one of the top ten "rising" sectors in EU's regional development plan. At this point, the EU's Blue Economy scope and classification and the annual reports it publishes (EC, 2017, 2018, 2019, 2020) are fundamental sources for the related literature.

The second important critical effort in developing the Blue Economy concept has been made by the UN and its related organisations. At the "Sustainable Development (Rio+20) Conference," though the concept was not used directly, it was emphasised that the targets to be reached for a green and sustainable development through the spatial planning of oceans and seas economically and socially could be attained (Silver et al., 2015; Garland et al., 2019). It was also pointed out that, within this context, the Blue Economy has the potential to contribute to minimising inequality and poverty in developing countries or regions, generating more income from marine resources while preserving the ecological balance, reducing national debts, and moving towards a low-carbon production (United Nations Conference on Sustainable Development, 2014). In general terms, United Nations Conference on Sustainable Development (2014) and related organisations such as UNCTAD (2014), UNEP (2015), and World Bank (2017) regard the Blue Economy, differently from the EU, as a concept within the green economy framework of social justice, equality, and environmental sustainability and especially as a subject concerning Small Island Developing States and Least Developed Coastal Countries. Indeed, realising social inclusion and ecological sustainability of oceans and coastal regions is viewed as the primary function of "Blue Economy" according to UN's 2030 Sustainable Development Goals (World Bank, 2017).

The third crucial effort in the development of the concept has been made by the USA's National Ocean Economics Program (NOEP). NOEP, founded in 1999, takes all economic activities conducted in oceans, seas, and coastal regions into the scope of Blue Economy but uses the concepts such as "ocean economy "and "coastal economy" instead of "the Blue Economy" (Colgan, 2013; Garland et al., 2019; Graziano et al., 2019). NOEP, which defines ocean economy as an economy consisting of 6 main sectors and 23 industries (sub-sectors) using the sea directly or indirectly, represents both the first and the most important initiative in the sea-based economy's defining, classifying, and calculating its share in GDP (Colgan, 2013; Surís-Regueiro et al., 2013). The definition and methodology with both geographical and industrial dimensions on ocean/coastal economy developed by NOEP have been a source of inspiration for numerous institutions and researchers' classification and calculation on blue/coast/ocean economy field. However, it has been noticed that the USA-centred NOEP classification reduces the Blue Economy to a more limited number of sectors than the EU classification (Garland et al., 2019).

Consequently, it has been seen that the uncertainty about the definition of Blue Economy, basically a "spatial intervention" for maximising the economic benefits of marine resources, continues (EC, 2012, 2020; Morrissey and O'Donoghue, 2012; Colgan, 2013; Surís-Regueiro et al., 2013; Zhao et al., 2014; Ketels and Protsiv, 2017; World Bank, 2017; Graziano et al., 2019), despite the considerable accumulation in literature in a short time and the methodological efforts to define and classify ocean/Blue Economy for the last decade (Winder and Le Heron, 2017; Voyer et al., 2018; Garland et al., 2019). In the literature, it is determined that the Blue Economy or blue growth strategies of the EU and the USA focus on improving the economy further in order to increase sectoral recovery and employment before the effects of the crises of various scales in the last quarter century deepen; however, in recent years, it has been determined that the UN strategy focuses on more sustainable, equitable, and ecologically balanced policy proposals (McKinley, 2022).

Upon the consideration of existing empirical literature on the Blue Economy, it has been seen that these researches are mostly focused on maritime/blue clusters (e.g. Doloreux, 2017; Ketels and Protsiv, 2017; Graziano et al., 2019; Shi et al., 2020) and sustainable development in economic dimension (e.g. Pinto et al., 2015; Heidkamp and Morrissey, 2019), the firms' strategy and policy application proposals in different spatial scales (Katila et al., 2019; Schultz-Zehden et al., 2019) through new knowledge generation, innovation, and smart specialisation (e.g. De Vet et al., 2016; Soma et al., 2018). There are also studies focusing on a sustainable environment, ecology, justice, and equality (e.g. Brent et al., 2020; Ertör and Hadjimichael, 2020), the viewpoints of the shareholders in seas and oceans (e.g. Howard, 2018), and governance and spatial planning (e.g. Ehlers, 2016; Choi, 2017). In addition to these, there are empirical studies on sustainable sea tourism within the scope of Blue Economy (e.g. Rogerson and Rogerson, 2019; Mayén Cañavate et al., 2019) and global-scale fishery and its sustainability (e.g. Pauly, 2018).

Upon a critical analysis of the literature, it could be noticed that the role of the Blue Economy in nation states' economic magnitudes such as growth, employment, productivity, and added value and its contribution to GDP is not known yet. It is also notable that the number of studies on sectoral dynamics and "economic performance" is not sufficient and that there is a considerable research gap in the field. As for Turkey, to the best of our knowledge, there is no empirical research revealing the potential of Turkey in the Blue Economy field, the magnitude, and the performance of the industries forming the Blue Economy at the national or regional levels. There is a huge gap in this field and this study is the first important step to fill this gap.

7.3 Data and Method

As highlighted earlier, the main aim of this study is to determine the size and performance of the Blue Economy in Turkey and to map the spatial distribution of Blue Economy industries. In this context, a classification that constitutes the Blue Economy and its sub-sectors is needed to determine the size of the Blue Economy within all economic activities and draw the sectoral cluster maps of the industries based on it. Although there have been several attempts to define and classify the structure of the maritime/Blue Economy in the last 10 years (EC, 2012, 2020; Colgan, 2013; Surís-Regueiro et al., 2013; Ketels & Protsiv, 2017; World Bank, 2017; Graziano et al., 2019), there is not yet an agreed standard on the definition and classification of the Blue Economy. In this study, the classification developed by the EU for the Blue Economy is based on economy as this classification is the most up-to-date and comprehensive. In addition, the fact that Turkey is close to the EU and in the process of integration has also been effective in the use of this classification (EC, 2018). The classification of the EU is important in that it focuses on "regional economic growth" rather than environmental sustainability (Garland et al., 2019) and reveals the sectors and clusters on which the blue sectors depend (Table 7.1).

As seen in Table 7.1, according to the EU classification, the Blue Economy has 6 main sectors, 10 sub-sectors or industries, and according to NACE Rev. 2,[3] it consists of 31 activities. The main limitation of the study is that some economic activities within the classification can be clearly defined as *exclusively maritime* (such as fishing and maritime transport), while the NACE classification for other sectors includes both land and maritime activities (such as cargo handling, storage, oil, and gas extraction), which are *partly maritime*. For this reason, the "maritime rate" of these activities should be estimated and calculated so that these activities can be fully evaluated within the Blue Economy. All other researchers have encountered this problem in defining the term. Therefore, some categorised the blue industries into two as "direct maritime" and "indirect maritime" industries (Morrissey & O'Donoghue, 2012) while others classified them as "exclusively maritime," "mostly maritime," and "partly maritime" industries (Surís-Regueiro et al., 2013). Colgan (2013) made a classification as "primarily or exclusively maritime" and as "coastal/sea/lake geographically." In addition, these researchers state that the indirect, mostly, or partly maritime criteria used

TABLE 7.1 Classification developed by the European Commission in 2018 for the Blue Economy

Sectors/Clusters	*Sub/Sector/Industry*	*Maritime proportion (%)*	*NACE code*	*Blue Economy activities within NACE Rev. 2 Code 4-Digit*
Marine Living Resources	Primary Production	100	03.10	Capture fisheries
		100	03.20	Aquaculture
	Processing and Distribution of Fish Production	100	10.20	Processing and preserving of fish, crustaceans and molluscs
		All provinces and 20%	46.38	Wholesale of other food, including fish, crustaceans, and molluscs
		100	47.23	Retail of fish, crustaceans, and molluscs in specialised stores
Marine Extraction of Oil and Gas	Oil and Gas	Excluded	06.10	Extraction of crude petroleum
		Excluded	06.20	Extraction of natural gas
		Excluded	09.10	Support activities for petroleum and natural gas extraction
Ports Activities	Ports	100	52.22	Service activities incidental to water transportation
		Coastal Provinces and 70%	52.24	Cargo handling
		Coastal Provinces and 70%	52.10	Warehousing and storage
	Construction of Water Projects	100	42.91	Construction of water projects
Shipbuilding and Repair	Shipbuilding	100	30.11	Building of ships and floating structures
		100	30.12	Building of pleasure and sporting boats
		100	33.15	Repair and maintenance of ships and boats
Maritime Transport	Shipping and Transport	100	50.10	Sea and coastal water passenger transport
		100	50.20	Sea and coastal water freight transport
		100	50.30	Inland passenger water transport
		100	50.40	Inland freight water transport
		100	77.34	Renting and leasing of water transport equipment

(Continued)

TABLE 7.1 Classification developed by the European Commission in 2018 for the Blue Economy *(Continued)*

Sectors/Clusters	*Sub/Sector/Industry*	*Maritime proportion (%)*	*NACE code*	*Blue Economy activities within NACE Rev. 2 Code 4-Digit*
Coastal Tourism	Accommodation	All of the coastal provinces	55.10	Hotels and similar accommodation
		All of the coastal provinces	55.20	Holiday and other short-stay accommodation
		All of the coastal provinces	55.30	Camping grounds, recreational vehicle parks, and trailer parks
		All of the coastal provinces	55.90	Other accommodation
	Transport	Excluded	47.30	Retail sale of automotive fuel in specialised stores
		Excluded	49.10	Passenger rail transport, interurban
		Excluded	49.31	Urban and suburban passenger land transport
		Excluded	51.10	Passenger air transport
	Other Expenditures	Excluded	47.60	Retail sale of cultural and recreation goods in specialised stores
		Excluded	47.70	Retail sale of other goods in specialised stores
		Excluded	56.00	Food and beverage service activities

Source: EC (2018: 191–192 and authors)

in the classification of the Blue Economy sectors may lead to the problem of overestimating or underestimating the value, size, and scope of the Blue Economy. Colgan (2013) argues that the coastal tourism sector and the marine renewable energy sector are the most difficult sectors to predict. One solution to overcome these problems is to exclude problem industries from cluster analysis as Graziano et al. (2019) did or to include them in the analysis under certain assumptions. Since we could not overcome the data/estimation problem[4] in calculating the marine ratio in this study, we used the approach of EC (2018) and also the methods of Surís-Regueiro et al. (2013) and Colgan (2013) and thus included the sectors that are completely marine and mainly marine as well as geographically marine (i.e. coastal provinces) using "maritime rate" criteria. We excluded the sectors (i.e. transport and other expenditures sub-sectors within coastal tourism) that are both partly marine and have no data for estimation of their maritime rate. In addition, since there is no gas and oil extraction from the sea in Turkey, and therefore there are no data, these two activities were excluded from the study as well.

In this quantitative study, the sectors, and activities within the scope of the Blue Economy in Turkey were defined according to the NACE Rev. 2 classification, and the microdata in the NUTS 3 (81 provinces) level covering the year 2018 were obtained from the TurkStat through the aggregation method. The microdata were calculated in terms of employment and value-added parameters according to 21 NACE economic activities with 4 digits within the scope of the Blue Economy. Although all data sources are from the micro data set of TurkStat, the primary production data covering only capture fisheries and aquaculture has been compiled from the Social Security Institution of Turkey since it is not included in the micro data set of TurkStat.

7.4 Results

7.4.1 Placing the Blue Economy in Turkey

Blue Economy accounted for 1.1% of the national GVA (gross value added) of Turkey and 1.4% of employment in the country in 2018 (Table 7.2). When compared with the EU averages, it is seen that the share of the Blue Economy in the total employment is 2.2% in the EU countries, while it has a share of 1.5% in the total added value (EC, 2020). Thus, the contribution of the Blue Economy in Turkey is slightly below the EU average (EC, 2020). However, its share in national economies varies between approximately 1–4% in many countries of the world. For example, while it constitutes 1.7% of GDP and 1.9% of employment in the USA (Colgan, 2013), this rate is 4.03% in China (Zhao et al., 2014). On the other hand, in terms of the total size of the Blue Economy, Turkey is the fifth in employment among 28 EU countries after Spain, the UK, Germany, and Italy and the seventh as far as the value-added. Given that Turkey is the sixth-largest economy in Europe in terms of GVA, it can be stated that the Blue Economy of the country is close to the EU level in terms of size and development level.

TABLE 7.2 Blue Economy in Turkey in 2018

Blue Economy and its sectors/clusters	*Number of companies*	*Persons employed*	*GVA ($)*	*Personnel costs ($)*	*Turnover ($)*
Marine living resources	3936	19,631	385,716,553	132,801,157	2469,336,903
Marine extraction of oil and gas	–	–	–	–	–
Port activities	2616	59,039	2189,396,030	791,191,573	5967,309,014
Shipbuilding and repair	2750	43,843	718,034,292	387,285,089	2580,285,310
Maritime transport	3960	29,111	1082,970,674	380,014,995	4404,074,891
Coastal tourism	16,692	245,785	4,805,047,894	1913,456,489	10,170,424,442
Blue Economy total	29,954	393,494	9181,165,442	3604,749,303	25,591,430,560
National economy rotal[5]	3160,392	28,738,000	778,382,000,000	121,280,751,415	1623,464,261,945
Blue Economy as % of national economy	0.95%	1.37%	1.18%	2.97%	1.58%

Source: Authors calculations based on data from TurkStat (2020).

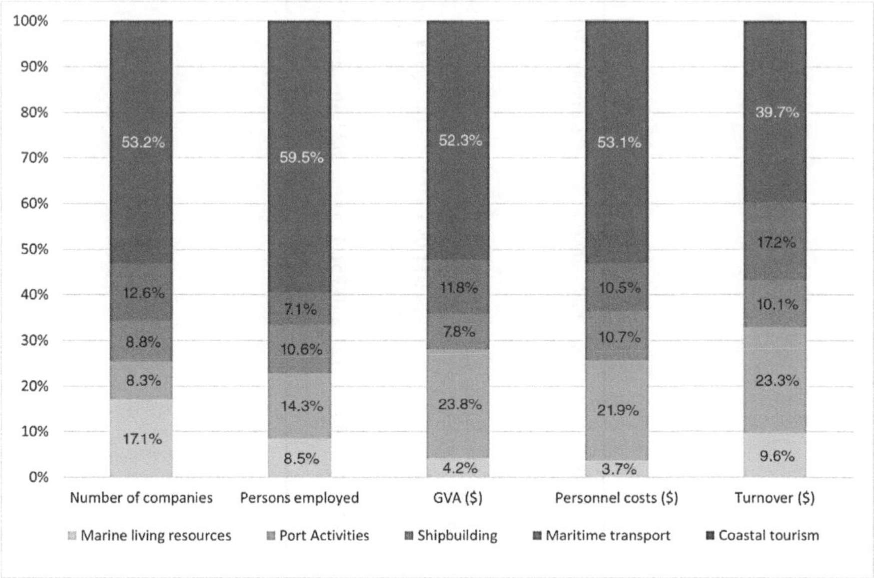

FIGURE 7.1 Economic performance of the Blue Economy sectors/clusters in Turkey, 2018 (Created by authors based on data from TurkStat, 2020)

Table 7.2 and Figure 7.1 show the distribution of the Blue Economy sub-sectors in terms of the number of companies, employment, GVA, turnover, and average wages in 2018. It is seen in the table and figure that the coastal tourism sector has considerable dominance. Coastal tourism accounts for more than 50% of the Blue Economy in all parameters, except turnover. These rates are not exclusive to Turkey. Coastal tourism dominates the Blue Economy in the EU countries (EC, 2020), the United States (Colgan, 2013), and China (Zhao et al., 2014). Tourism plays an important role in the national economy and the funding of the foreign trade deficit in Turkey, which is among the important tourism destinations. The share of coastal tourism in Turkish economy is higher than it was in the economies of countries such as Spain and Greece. In this study, the data used to estimate coastal tourism in the EU and the ratios that enable technical computing could not be obtained; thus, we stipulate that the values related to coastal tourism in Turkey have been estimated lower than they actually are. As Colgan (2013) maintains, coastal tourism is the sector that is most problematic and difficult to estimate within Blue Economy.

Port activities are the second most important contributor to the Blue Economy in Turkey (Figure 7.1). In general, about 20% of the Blue Economy-based jobs in Turkey are related to port shipping and warehousing. Port activities have a very important place in Turkey, which is surrounded by the Mediterranean, Aegean Sea, and the Black Sea. Although the number of companies is small compared to the EU, port activities have a performance above the EU average in terms of employment, GVA, wage, and turnover. Port activities accounted for 11% of employment and 16% of GVA in Blue Economy in the EU in 2018. Among 28 blue economies in the EU, Turkey is the third-largest economy in terms of port activities.

In Turkey, while the third important sector within the Blue Economy in terms of the value-added is maritime transport, it is the shipbuilding industry based on the employment parameters (Figure 7.1). The maritime transport sector, which is superior to shipbuilding in terms of GVA as well as the number of companies and turnover, plays an important role in Turkey's economy. While maritime transport accounts for 8% of employment and 16% of GVA in the EU Blue Economy, Turkey remains below the EU average and ranks third in Europe after Germany and Italy (Figure 7.1; EC, 2020).

The shipbuilding industry, which is another Blue Economy sector, includes the construction, maintenance, and repair of ships, boats, and yachts and has an important share in terms of employment and added value in Turkey (Figure 7.1). Turkey, which is close to the employment and added value share of this highly innovative and competitive sector of the Blue Economy (6% and 8%) in the EU, follows the UK, which is the EU leader in the sector (EC, 2020). The shipyards in Turkey are among the leading countries in terms of ship and boat manufacturing, maintenance, and repair of ships after the European shipyards, which are leaders in cruise ship manufacturing, and after the Asian shipyards such as China and South Korea, which are the world leaders in the cargo tanker industry.

The marine living resource sector involving fisheries and aquaculture as well as the processing and distribution of fish products has the lowest employment and value-added in Turkey, as expected since it is mostly based on primary activities. Considering its performance in EU countries (12% and 18%), where Spain is the leader, Turkey ranks ninth in Europe. Turkey, which is surrounded by seas, is far behind the EU in terms of fisheries and the production of fish products. This is because the employment and GVA of the sector cannot be measured due to the lack of data and the sector has some structural and technological problems.

Finally, although in recent years Turkey has accelerated its oil and gas exploration activities in the sea, there has yet been no discovery operated commercially. Therefore, there are no data available regarding this sector in the Blue Economy. However, in EU countries, it is seen that this sector has a share of 1% in employment and a share of 9% in GVA (EC, 2020).

7.4.2 Regional Distribution of Blue Economy

The spatial dimensions of the Blue Economy have also been analysed at the regional level in absolute and relative terms. When the spatial pattern in Figures 7.2 and 7.3 is examined, it is seen that Blue Economy is geographically concentrated almost entirely in the coastal areas. While 28 coastal provinces constitute 95% of the Blue Economy, the share of 53 inland provinces is only 5%.[6] While the number of employees in the Blue Economy per thousand people is nine in the coastal areas, it is one in the inner regions. In coastal areas, the added value per capita in Blue

FIGURE 7.2 Share of Blue Economy employment by provinces in Turkey, 2018

(Created by authors based on data from TurkStat, 2020)

Economy is $194, whereas it is $13 in the inland provinces. It is also observed that 77% of the employment in the Blue Economy in Turkey (Figure 7.2) and 85% of the added value is concentrated in five major metropolitan cities in the west (Istanbul, Antalya, Muğla, Izmir, and Ankara), which also lead the industrial production and service activities in Turkey. When five other metropolitan cities, namely, Kocaeli, Mersin, Bursa, Aydın, and Balıkesir, are also added to these five cities, it is seen that 88–95% of the Blue Economy runs in the first ten provinces (Figure 7.2). While Istanbul, where the population, production, and trade are the most concentrated in Turkey, constitutes 38–46% of the Blue Economy alone, Antalya ranks second with 22% in relation to the intensification of tourism and yacht-building activities. Muğla and İzmir are places where tourism, fishing, and processing of fish products are concentrated, while Kocaeli and Bursa provinces are among the places where the manufacturing industry is concentrated in the west of Turkey. In addition, Istanbul, Izmir, and Mersin are the places where port activities are concentrated, as they are Turkey's gateway to the world. The high share of the Capital City, Ankara is related to the specialisation on water projects. This criterion, which shows the "absolute size" and potential of the Blue Economy on a regional/provincial basis, reveals that places with a large population and economy naturally have a larger Blue Economy.

Another alternative criterion that can be used to evaluate the Blue Economy based on region/province is to consider the "relative size." In this context, when the regional Blue Economy is normalised according to the population of the provinces, the advantages arising from size disappear and the real performance of the provinces is revealed. Accordingly, while coastal provinces are the leading ones in terms of Blue Economy employment per thousand people, important changes occur in the ranking of provinces. While Istanbul falls from leadership to fifth place in employment (Figure 7.3), Antalya rises from second to first place followed by Muğla and Yalova. This is related to tourism activities in the context of the development of suitable climate and infrastructure opportunities in Antalya and Muğla, and the intensification of the shipbuilding industry in

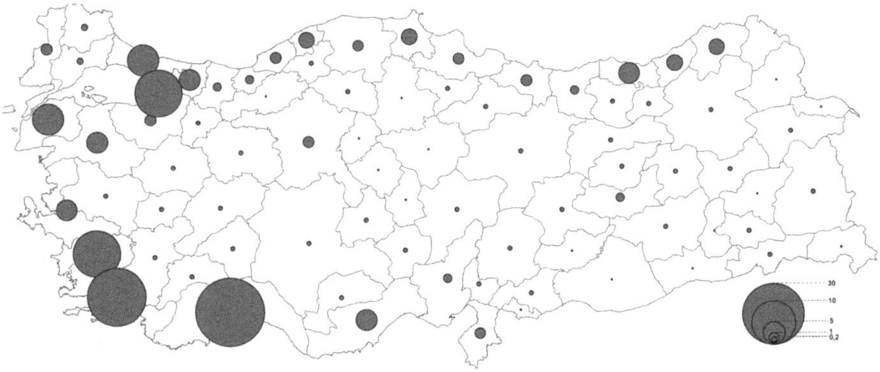

FIGURE 7.3 Spatial distribution of Blue Economy employment per thousand people by provinces in Turkey in 2018

(Created by authors based on data from TurkStat, 2020)

Yalova. The shipbuilding industry has developed not only in Yalova but also in Istanbul (Tuzla) and Kocaeli (Gölcük), which are located in a triangle of 50 km^2 in the immediate vicinity. İzmir moves from 4th place to 11th place in employment. The capital Ankara (Figure 7.2), which was in 5th place earlier, falls to the 17th (Figure 7.3), especially due to the increase in administrative activities. It is noteworthy that the Blue Economy is generally concentrated in the provinces in the socio-economically developed west of the country and the regions on the shores of the Mediterranean, Aegean, and Marmara seas. The longest coasts of the country and the most extensive activities of fisheries and aquaculture are on the Black Sea coasts in the north and northeast; however, the provinces along this coast (except Trabzon) are not in the top ten in neither absolute nor relative size. Because the most important economic activity on the Black Sea coast is agriculture and service, these results indicate that, whether it is absolute or relative, the spatial distribution of the Blue Economy in Turkey shows parallelism in terms of employment and the value-added.

7.5 Conclusion

This study examines the Blue Economy, which offers important opportunities for stakeholders in the context of sustainable development in the case of Turkey. Using the European Commission's (EC) Blue Economy framework, this research, which does not focus on social and cultural dynamics, aims to reveal the place of the Blue Economy within the Turkish economy. In addition, this study has displayed the share of the provinces in the country's Blue Economy and their spatial distribution on a regional scale.

According to the findings of the study, it was found out that, first of all, the Blue Economy has a 1–1.5% share in the national economy of Turkey. Turkish Blue Economy employs roughly half a million people and approximately generates a GVA of 10 billion dollars. It has been also found that Coastal tourism dominates the Turkish Blue Economy as it does in the rest of the world (Colgan, 2013; Zhao et al., 2014; EC, 2020). This is followed by port activities and shipbuilding (EC, 2020), while maritime transport (Morrissey and O'Donoghue, 2013) and living resource sectors/clusters are the last in employment and value-added, respectively. On the other hand, activities related to deep-sea mining and marine energy generation, which are the Blue Economy sectors, are non-existent in Turkey. In order to increase welfare and employment, in Turkey, where oil and gas extraction activities from the sea have just begun, economic growth should be achieved without harming the ecosystem.

In addition to the fact that a large part of the population lives in the coastal areas in Turkey, especially the manufacturing industry and service activities are concentrated in the provinces in the west of the country. Thus, the Blue Economy has spatially concentrated almost entirely along the coast and also in the western regions. Coastal regions form 95% of it while the hinterland's share is only 5%. Another finding is that a very large part (i.e. 80% of it) of the Blue Economy in Turkey is concentrated in five big cities (İstanbul, Antalya, Muğla, İzmir, and Ankara) in the West, also leaders in the production of industry and service in the country. These findings reveal that the Blue Economy has concentrated in the developed Western and coastal regions of the country (the only exception is the Capital Ankara in the interior part). However, this showed that blue growth in Turkey is economy oriented as in the EU framework, and that Blue Economy activities are not evenly distributed among the regions of the country.

In brief, this study has focused on the economic part of the Blue Economy rather than the social dimension as stated by McKinley (2022). Therefore, it examines the Blue Economy at a national and spatial scale for the sustainable development of Turkey for the first time by filling an important gap in the literature. The central government needs to work together with the regional stakeholders on the Blue Economy, which has a significant potential in Turkey's economic development (Pinto et al., 2015; Soma et al., 2018), in order to develop more sustainable (Surís-Regueiro et al., 2013; Ehlers, 2016; EC, 2019, 2020), inclusive, and equitable social justice (McKinley, 2022) and governance policies (Graziano et al., 2019; Heidkamp and Morrissey, 2019) without harming the marine ecosystem. In this context, for example, the "blue flag" application for environmental protection on the beaches of the Turkish coast is important to demonstrate the sustainability of both the sea life and tourism sector. However, while this was not considered in this study, it has the potential to be an important research topic for future studies. On the other hand, societies have significant opportunities to realise a sustainable and fairer Blue Economy in the coming years.

Notes

1 In addition to the Blue Economy concept, in literature, both researchers and international organizations use "ocean economy," "marine economy," "coastal economy," "blue growth," and "blue cluster" sometimes in the same meaning and sometimes with a different content.

2 The efforts of the EU in Blue Economy area, in fact, dates back to 2007. "Blue Book" prepared by the EU in 2007 is one of the framework documents including the integrated governance and planning of the marine policy of the union (Surís-Regueiro et al., 2013).
3 NACE is the EU equivalent of NAICS, the North American Industry Classification System. Rev. 2 is the latest revised version of these codes.
4 In order to overcome this data problem, the EU has used a wide variety of techniques and calculations. In cooperation with the institutions within the EU, it calculated and estimated the maritime rates of mostly and partly marine activities. Unfortunately, this is not possible in our study due to lack of such data.
5 Sectors given constitute coverage of TurkStat Annual Industry and Service Statistics included in NACE Rev.2 Classification in the European Community. However, Section A, K, O, T, U, and S 94 division in the NACE Rev.2 is not covered in TurkStat Annual Industry and Service Statistics due to data privacy (TurkStat, 2020). Therefore, the sectors mentioned above are not included in the number of companies, employment, value added, personnel costs, and turnover parameters used in this study.
6 For example, if the same thing (coastal vs. interior comparison) applied to the distribution of population in Turkey, this distribution would be as follows: 55% of the population live in coastal provinces, while 45% live in inland provinces.

References

Brent, Z. W., Barbesgaard, M., & Pedersen, C. (2020). The Blue Fix: What's driving blue growth? *Sustainability Science*, 15(1), 31–43. https://doi.org/10.1007/s11625-019-00777-7

Choi, Y. R. (2017). The Blue Economy as governmentality and the making of new spatial rationalities. *Dialogues in Human Geography*, 7(1), 37–41. https://doi.org/10.1177%2F2043820617691649

Colgan, C. S. (2013). The ocean economy of the United States: Measurement, distribution, & trends. *Ocean & Coastal Management*, 71, 334–343. https://doi.org/10.1016/j.ocecoaman.2012.08.018

De Vet, J.-M., Edwards, J., & Bocci, M. (2016). Blue growth and smart specialisation: How to catch maritime growth through 'Value Nets'. *S3 Policy Brief Series No. 17/2016*. Seville: Joint Research Centre. file:///Users/ragipkalelioglu/Downloads/jrc100975_pb_bg_and_s3_final%20(1).pdf (Accessed 20 December 2020).

Doloreux, D. (2017). What is a maritime cluster? *Marine Policy*, 83, 215–220. https://doi.org/10.1016/j.marpol.2017.06.006

Doussineau, M., Gnamus, A., Gomez, J., Haarich, S., & Holstein, F. (2020). Smart specialisation and blue biotechnology in Europe. *JRC Science for Policy Report*. https://op.europa.eu/en/publication-detail/-/publication/c54d0861-4016-11eb-b27b-01aa75ed71a1/language-en (Accessed 01st December 2020).

Ehlers, P. (2016). Blue growth and ocean governance—How to balance the use and protection of the seas. *WMU Journal of Maritime Affairs*, 15, 187–203. https://doi.org/10.1007/s13437-016-0104-x

Ertör, I., & Hadjimichael, M. (2020). Blue degrowth and the politics of the sea: Rethinking the Blue Economy. *Sustainability Science*, 15(1), 1–10. https://doi.org/10.1007/s11625-019-00772-y

European Commission (EC). (2012). *Blue Growth. Opportunities for Marine and Maritime Sustainable Growth*. https://ec.europa.eu/maritimeaffairs/sites/maritimeaffairs/files/docs/publications/blue-growth_en.pdf (Accessed 15th August 2020).

European Commission (EC). (2017). *Commission Staff Working Document: Report on the Blue Growth Strategy: Towards More Sustainable Growth and Jobs in the Blue Economy.* https://ec.europa.eu/maritimeaffairs/sites/maritimeaffairs/files/swd-2017-128_en.pdf (Accessed 15th August 2020).

European Commission (EC). (2018). *The 2018 Annual Economic Report on the EU Blue Economy.* https://op.europa.eu/en/publication-detail/-/publication/79299d10-8a35-11e8-ac6a-01aa75ed71a1 (Accessed 15th August 2020).

European Commission (EC). (2019). *The EU Blue Economy Report 2019.* https://op.europa.eu/en/publication-detail/-/publication/676bbd4a-7dd9-11e9-9f05-01aa75ed71a1/language-en/format-PDF/source-98228766 (Accessed 15th August 2020).

European Commission (EC). (2020). *The EU Blue Economy Report 2020.* https://ec.europa.eu/maritimeaffairs/sites/maritimeaffairs/files/2020_06_blueeconomy-2020-ld_final.pdf (Accessed 15th August 2020).

Fisheries and Oceans Canada. (2020). *Fisheries and Oceans Canada.* https://www.dfo-mpo.gc.ca/index-eng.html (Accessed 5th October 2020).

Garland, M., Axon, S., Graziano, M., Morrissey, J., & Heidkamp, C. P. (2019). The Blue Economy: Identifying geographic concepts and sensitivities. *Geography Compass*, 13(7), e12445. https://doi.org/10.1111/gec3.12445

Graziano, M., Alexander, K. A., Liesch, M., Lema, E., & Torres, J. A. (2019). Understanding an emerging economic discourse through regional analysis: Blue Economy clusters in the US Great Lakes basin. *Applied Geography*, 105, 111–123. https://doi.org/10.1016/j.apgeog.2019.02.013

Heidkamp, C. P., & Morrissey, J. (Eds.). (2019). *Towards Coastal Resilience and Sustainability.* New York: Routledge.

Howard, B. C. (2018). Blue growth: stakeholder perspectives. *Marine Policy*, 87, 375–377.

Karahan, C. B., & Kirval, L. (2018). Clustering potential of Istanbul maritime sector. *Maritime Business Review*, 3(3), 314–336. https://doi.org/10.1108/MABR-03-2018-0012

Katila, J., Ala-Rämi, K., Repka, S., Rendon, E., & Törrönen, J. (2019). Defining and quantifying the sea-based economy to support regional blue growth strategies—Case Gulf of Bothnia. *Marine Policy*, 100, 215–225. https://doi.org/10.1016/j.marpol.2018.11.035

Ketels, C., & Protsiv, S. (2014). *Methodology and Findings Report for a Cluster Mapping of Related Sectors.* https://ec.europa.eu/docsroom/documents/16527/attachments/1/translations/en/renditions/native (Accessed 10th August 2020).

Ketels, C., & Protsiv, S. (2017). *Priority Sector Report: Blue Growth, European Cluster Observatory Report.* https://ec.europa.eu/docsroom/documents/24681/attachments/2/translations/en/renditions/pdf. (Accessed 10th August 2020).

Mayén Cañavate, B., Bernal Conesa, J. A., Briones Peñalver, A. J., & Anunciação, P. (2019). Tourism in the Blue Growth strategy: A model proposal. *Anatolia*, 30(2), 267–278. https://doi.org/10.1080/13032917.2018.1519228

McKinley, E. (2022). A Blue Economy for who? Linking marine social sciences with the Blue Economy discourse. In: P. Heidkamp and J. Morrissey (Eds.), *Blue Economy: People and Regions in Transitions.* New York: Routledge.

McKinley, E., Acott, T., & Yates, K. L. (2020). Marine social sciences: Looking towards a sustainable future. *Environmental Science & Policy*, 108, 85–92. https://doi.org/10.1016/j.envsci.2020.03.015

Morrissey, K., & O'Donoghue, C. (2012). The Irish marine economy and regional development. *Marine Policy*, 36(2), 358–364. https://doi.org/10.1016/j.marpol.2011.06.011

Morrissey, K., & O'Donoghue, C. (2013). The role of the marine sector in the Irish national economy: An input–output analysis. *Marine Policy*, 37, 230–238. https://doi.org/10.1016/j.marpol.2012.05.004

National Oceanic and Atmospheric Administration (NOAA). (2018). *NOAA Report on the U.S. Ocean Marine Economy*. http://coast.noaa.gov/digitalcoast/training/econreport.html (Accessed 5th October 2020).

OECD. (2016). *The Ocean Economy in 2030*. Paris: OECD Press.

Our Ocean Health. (2020). *Harnessing Our Ocean Wealth—An Integrated Marine Plan for Ireland*. https://www.ouroceanwealth.ie/about-us (Accessed 5th October 2020).

Pauly, D. (2018). A vision for marine fisheries in a global Blue Economy. *Marine Policy*, 87, 371–374.

Pinto, H., Cruz, A. R., & Combe, C. (2015). Cooperation and the emergence of maritime clusters in the Atlantic: Analysis and implications of innovation and human capital for blue growth. *Marine Policy*, 57, 167–177. https://doi.org/10.1016/j.marpol.2015.03.029

Rogerson, C. M., & Rogerson, J. M. (2019). Emergent planning for South Africa's Blue Economy: Evidence from coastal and marine tourism. *Urbani Izziv*, 30, 24–36. https://doi.org/10.5379/urbani-izziven-2019-30-supplement-002

Schultz-Zehden, A., Weig, B., Lukic, I. (2019). Maritime spatial planning and the EU's blue growth policy: Past, present and future perspectives. In: J. Zaucha and K. Gee (Eds.), *Maritime Spatial Planning* (pp. 121–149). Cham: Palgrave Macmillan.

Shi, X., Jiang, H., Li, H., & Xu, D. (2020). Maritime cluster research: Evolutionary classification and future development. *Transportation Research Part A: Policy and Practice*, 133, 237–254. https://doi.org/10.1016/j.tra.2020.01.015

Silver, J. J., Gray, N. J., Campbell, L. M., Fairbanks, L. W., & Gruby, R. L. (2015). Blue Economy and competing discourses in international oceans governance. *The Journal of Environment & Development*, *24*(2), 135–160. https://doi.org/10.1177%2F1070496515580797

Soma, K., van der Burg, S. W. K., Hoefnagel, E., & Stuiver, M. (2018). Social innovation—A future pathway for blue growth? *Marine Policy*, 87, 363–370. https://doi.org/10.1016/j.marpol.2017.10.008

Surís-Regueiro, J. C., Garza-Gil, M. D., & Varela-Lafuente, M. M. (2013). Marine economy: A proposal for its definition in the European Union. *Marine Policy*, *42*, 111–124. https://doi.org/10.1016/j.marpol.2013.02.010

TurkStat. (2020). *Annual Industry and Service Statistics: Micro Data Set 2018*. Ankara: Turkish Statistical Institute.

United Nations Conference on Sustainable Development. (2014). *Blue Economy Concept Paper*. https://sustainabledevelopment.un.org/content/documents/2978BEconcept.pdf (Accessed 28th December 2020).

UNCTAD. (2014). *The Oceans Economy: Opportunities and Challenges for Small Island Developing States*. https://unctad.org/system/files/official-document/ditcted2014d5_en.pdf (Accessed 28th December 2020).

UNEP. (2015). *Blue Economy: Sharing Success Stories to Inspire Change*. https://www.unep.org/resources/publication/blue-economy-sharing-success-stories-inspire-change (Accessed 28th December 2020).

Voyer, M., Quirk, G., Mcilgorm, A., & Azmi, K. (2018). Shades of blue: What do competing interpretations of the Blue Economy mean for oceans governance? *Journal of Environmental Policy and Planning*, 20(5), 595–616. https://doi.org/10.1080/1523908X.2018.1473153

Winder, G. M., & Le Heron, R. (2017). Assembling a Blue Economy moment? Geographic engagement with globalizing biological-economic relations in multi-use marine environments. *Dialogues in Human Geography*, 7(1), 3–26. https://doi.org/10.1177%2F2043820617691643

World Bank. (2017). *The Potential of the Blue Economy*. https://openknowledge.worldbank.org/bitstream/handle/10986/26843/115545.pdf?sequence=1&isAllowed=y (Accessed 28th December 2020).

Zhao, R., Hynes, S., & He, G. S. (2014). Defining and quantifying China's ocean economy. *Marine Policy*, 43, 164–173. https://doi.org/10.1016/j.marpol.2013.05.008

8

EXPERIMENTATION AND ENACTIVE RESEARCH

Building a Knowledge Infrastructure for Marine Social Science

Nicolas Lewis and Richard Le Heron

8.1 Introduction

In this chapter, we argue that the work of critical social scientists in Aotearoa New Zealand's Sustainable Seas National Science Challenge (SSNSC) has been instrumental in *mobilising the "Blue Economy moment"* to build a knowledge infrastructure for enacting novel environment-economy relations. The SSNSC is a ten-year, $NZ71 million research programme that was given the objective of "enhancing the use of New Zealand (NZ) marine resources within environmental and biological constraints." It was also given an unprecedented mandate to disrupt business as usual in the production of governmental knowledge of the oceans. We examine how SSNSC has used this mandate to translate its objective into *a political project of knowledge production* (Lewis and Le Heron 2019) and the creation of a new *knowledge infrastructure* (Edwards 2017) for *doing marine economy differently*. Its work has begun to shift investment frameworks, policy imaginaries and practices, and public engagements with ocean futures.

The chapter details how NZ social scientists have established a community that is now well-networked into transformative initiatives and governmental agencies and has begun to secure a place for marine social science in research and practice. The case is testament to the value of calls by McKinley et al. (2020), Bennett (2019), and others to create such communities. In particular, it shows how a social science community can emerge from research practices that engage actively in making other worlds possible (Roelvink et al. 2015). While the SSNSC case has defining spatio-temporal specificities, we draw insights from it that will support others to take up McKinley and colleagues' call to action. In detailing what has worked, however, we go further to summon a novel social science that avoids the twin traps of a science-inflected advocacy for social science and a "declarative" stance. We challenge social scientists to *embrace a more "enactive"*

DOI: 10.4324/9781003280248-11

social science focused on the possibilities of transitioning to new environment-economy relations. In this way, we extend McKinley (2023)'s reflection (Chapter 2, this volume) on the potential of *Blue Economy thinking* to show how it can make change in messy empirical worlds by escaping pre-defined categories of knowledge, social action, and economy.

8.2 Enactive Research as Methodology and Praxis

The chapter examines how SSNSC has been performed in practice to encompass significant levels of applied and critical social science. To do this, we draw on participant observation, public documents, published papers, and five years of insider conversations as active participants and strategic brokers in the development of SSNSC. The authors are two geographers who both led core research projects within SSNSC and participated in others. Richard was a core member of the team that established SSNSC, while by the end of Phase I Nick had become a member of the SSNSC leadership team. Richard wrote a foundational article on the emergence of Blue Economy discourse in the global arena (Winder and Le Heron 2017a), while Nick led Aotearoa New Zealand's first blue economy themed project. We have been strategically positioned within SSNSC and have worked together for more than 15 years building networked knowledge production projects in NZ social science (Le Heron and Lewis 2011).

In this chapter, we follow ourselves as disruptive knowledge framers through the design and delivery of SSNSC. This approach is not exactly ethnography, but nor is it simply ex-post reflection. Rather, we interrogate our own unprojected and performative knowledge making. Elsewhere, we have represented and theorised this approach as a wild-leaning praxis, which we have labelled "enactive research" (Carolan 2013; Lewis et al. 2016). The core components of this approach are a disposition to experiment with novel analytical categories in encounters with research participants, treat research encounters as opportunities to create impact directly, and reinterpret the academic as part of, rather than external to, collective social agency.

8.3 New Zealand's National Science Challenges

New Zealand's National Science Challenges were designed in the early 2010s as platforms for consolidating national scientific research into strategic, long-term responses to "issues that matter to all New Zealanders." Each was to disrupt existing scientific practice and to fracture distinctions between applied and fundamental science, investigator-led and mission-led science, and contestable and core long-term funding. The challenges were expected to establish collaborative practices of knowledge making. They were given three clear directives. First, they were to be led by their mission-oriented objectives and answerable to stakeholder panels and governance boards (including corporate leaders). Second, their objectives were to include demonstrable contributions to a national economic development agenda centred on export-led growth and wrapped in a pervasive economic nationalism that gave them an ordering rationality, purpose, and legitimacy. Third, they were to embrace and uphold obligations under the Treaty of Waitangi to centre questions of Māori rights and interests in their work and organisation and incorporate the notion of *mātauranga Māori*, a mix of Māori world views (Te Ao Māori) and the indigenous knowledge and ways of knowing that underpin them (Rout et al. 2021). The Aotearoa New Zealand state is bound to consider the implications of regulatory and development initiatives for Māori (Ruru 2018), especially in relation to environmental decision-making. In the marine commons, these obligations are even more pronounced. The settlement of injustices with respect to historical breaches of the Treaty and the return of assets and resources has made iwi (Māori tribes) prominent, if not dominant, economic, and political actors (Reid et al. 2019). These various decolonising obligations recognise Māori rights and interests as defining dimensions of social and environmental realms and matters of methodological, theoretical, and empirical significance in environmental research.

8.4 Sustainable Seas: Preconditions for a Marine Social Science

The SSNSC was established under the management of government science laboratory NIWA (National Institute for Water and Atmospheric research) and carved out of networks of senior marine scientists across multiple science organisations. This group worked with stakeholders to develop a set of core projects and establish a contestable funding pool for other projects. It centred the research programme on the concept of ecosystem-based management (EBM) and developing the practical frameworks, institutions, and tools for securing and implementing it. In conceiving its projects, the group built on multiple scientific research trajectories with connections to marine management. Mission priorities focused attention on science-led economy, but the key actors were guided by shared commitments to securing the future health of coasts and oceans and delivering diverse values beyond narrow economic objectives. Māori interests and several state agency stakeholders focused attention on questions of management and governance.

The complex set of conditions of possibility for SSNSC thus included a mandate for disruption and reassembly, a growth imperative, commitments to Māori, a deep environmentalism, and an institutionalised emphasis on science. Despite the assumption of many within its stakeholder communities that its purpose was to fund science that supported growth within limits, SSNSC was always a more open framework for action, where the need for social science was clear. This did not immediately suggest critical theorists, but such a group had already begun to carve out a place for social science in ocean science networks by forging a new social-ecological research community (see Table 8.1). Lengthy timelines in establishing the national science challenges' facilitated access into SSNSC for this group, which included the authors.

TABLE 8.1 An emerging NZ marine social science

Date	*Initiative*	*Opportunity*	*Knowledge Production effect*
1999–06	Three PhDs	Build networks with ecologists	Dialogue between fisheries and ecosystem management
2011–14	Marine Futures	Build marine social science expertise Lead scenario-based workshops	Develop socio-ecological knowledge and enactive methodologies Blue Economy as emergent and relational
2014–15	Centre of Research Excellence (CoRE) bid	Craft a multi-disciplinary discovery-oriented ocean science programme	Deepened networks Social scientists in agenda setting roles
2015	Marine social science workshop	Dialogue among scientists and social scientists	Identify marine social scientists—establish rudiments of network
2015	Tranche 1 projects	Install social science in challenge	Research participatory approaches, debate values, define Blue Economy
2016	Project leaders' meeting	Social scientists critique science projects, take on leadership roles	Centre social science in SSNSC
2018	Blue Economy theme	Fund projects centred on just transitions to a Blue Economy	Centre environment-economy relations as knowledge object for enactive marine social science
2019–20	Review and Phase II projects	Centre Social Science in Challenge	Transdisciplinary knowledge—platform for enactive social science

Geographers from the University of Auckland (UoA) had been working on questions of marine policy for a number of years. Three PhDs were completed in the early 2000s (Le Heron et al. 2008), leading to research relationships with NIWA-based marine ecologists. These relationships paved the way for UoA geographers to take part in "Marine Futures" (MF), a major government-funded science project headed by NIWA ecologists in the early 2010s. Social scientists carried out the applied policy work against which the project was to be evaluated, running workshops, focus groups, and surveys to collect and collate values and preferences for modelling and calculating trade-offs. The UoA group pushed the boundaries of this role to fashion a social science domain around sustainable development pathways and practice-based exploration that engaged key actors in debates about socio-ecological knowledge in marine environments (Thrush et al. 2016). Drawing on the lessons of earlier enactive work with stakeholders (Lewis et al. 2016), we devised scenario experiments that supported participants to imagine future knowledge institutional formations for redirecting environment-economy relations towards more ecologically sustainable futures (Le Heron et al. 2016). The project also initiated engagement with notions of Blue Economy (Winder and Le Heron 2017a).

The MF project assembled a transdisciplinary team and established a socio-ecological research momentum. With marine ecologist colleagues, we built a socio-ecological research stream as part of a bid for a major multi-year, inter-institutional, marine environmental science bid for a Centre of Research Excellence (CoRE) in 2015. We wrote social science into the bid against some opposition from other marine scientists, introducing unfamiliar and disruptive elements of social knowledge production to the mix. These included participatory environmental management. The bid was unsuccessful but normalised a role for social research formed around a conception of socio-ecological knowledge and sustainability transitions. This was a significant step forward from a more reductive understanding of the role of social science as either measuring social impact or outlining social values as elements of human agency otherwise understood as individualised, predetermined, and rational (or irrational and requiring correction). The bidding process firmed up the networks formed under MF and created conditions for SSNSC.

Elsewhere, critics from multiple disciplines questioned a suite of social and environmental justice issues (Hamilton-Hart and Stringer 2016; Peart 2018; McCormack 2018). Their work focused renewed attention on the governance of the marine commons at a time when Māori claims on the commons and challenges to science practice were intensifying (Bargh 2014; Reid et al. 2019; Rout et al. 2021). These pressures created ground for an alliance of concerns with the emerging enactive social-cultural and social-ecological research of SSNSC social scientists (Le Heron E. et al. 2019). The final element of the cocktail was the poststructuralist insight that *economy is relational, made, and more than human* (Lewis et al. 2016). Encouraged by our experience with ecologists, our growing interest in Blue Economy, and SSNSC's mandate to be disruptive, we laid out a defined role for social science as *transforming how we do economy* (Lewis 2018). We began to narrate the primary challenge facing the oceans as *doing economy differently*, at multiple spatial, temporal, and institutional scales. This demanded a transformative approach to understanding economy and the practice of environmental management. Our place in already established networks positioned us to re-script expected approaches, refocus existing questions, and advocate for social science as potentially transformative research practice at the core of SSNSC (Le Heron E. et al. 2020).

8.5 Social Science in Sustainable Seas

Phase I of SSNSC was comprised of 40 projects composed around a set of core projects set within six themes and developed with input from Māori and stakeholders. While understood to be interdisciplinary, the themes and projects developed along lines defined by the knowledge bases required to answer core questions. Two of the programmes *Our Seas* and *Valuable Seas* became largely social science driven—concerned with cultivating economic values, accounting for social values, encouraging participatory processes, and managing ecological processes. A third, *Managed Seas* became a site for the development of biophysical-science-informed models and related tools for management, with some minimal supporting applied social science. The *Tangaroa* programme hosted locally focused Māori-interest projects that brought together social science and mātauranga Māori around Te Ao Māori concepts and principles (see above). Social scientists also led two cross-programme projects and four contestable projects. In total 17 projects were led by trained and practising social scientists, with at least five other social-science-focused projects. Across and beyond these projects, insights from social science informed SSNSC's approach to addressing environment-economy-society connections relationally. The largest programme, Dynamic Seas, was ecology-based but owed a debt to MF and its networks and included social scientists in many of its initiatives.

In all, the core projects included nine geographers, 15 other social scientists, and eight mātauranga Māori informed social scientists, making up nearly a third of SSNSC researchers. Several social scientists were involved in two or more projects and across two or more research programmes. A social science research community began to take form as the projects took shape and started to rework scientised forms of social science into the practice of a distinctive Aotearoa New Zealand marine social science. It began to cohere around this sense of its own potential within SSNSC's mission, as well as the shared practices and experiences of SSNSC processes (workshop methodologies, impact-based outputs, Challenge conferences and project leaders' meetings, and collaboration across disciplines and in relation to stakeholder and Māori interests). It also cohered around the notion of "*mountains to the seas" a social-ecological rescaling of objects and subjects of government that draws on Te Ao Māori to reframe environment-economy relations in holistic terms* that transcend compartmentalised land and sea institutions and knowledge categories (Winder and Le Heron 2017b). With a significant numerical presence and preparedness to operate in disciplinary and institutional borderlands, social scientists began to shape the direction of the wider SSNSC.

The authors of this chapter were most closely associated with *Our Seas* and *Valuable Seas*, which were founded on relational understandings of ecological, social, and economic interactions as a platform for transitioning to new practices. Research trajectories focused on core propositions: management and economy are part of the same set of processes and practices (Lewis 2018); the changes in our oceans are driven by human activities, including loss of biodiversity, modification of seascapes, and climate change (Ulrich et al. 2018); uses are contested and impacts are cumulative (Davies et al. 2018); values (cultural economic, social, and ecological) are multiple, contingent, dynamic, and interrelated (Šunde et al. 2018); and participatory processes of governance and management are crucial for just transitioning (Le Heron E. et al. 2021). This work translated concepts of "growth within limits" into new appreciation for marine resourcefulness, relational processes, and participatory management (Lewis et al. 2020, 2021a). In wrapping up much of this work at the completion of Phase I, we reinterpreted (and communicated) this message for public audiences as working in and with their oceans to generate intergenerational livelihoods and flourishing ocean ecologies (Le Heron E. et al. 2021).

Social scientists are more prominent in the leadership of Phase II of SSNSC. They made up three of the eight-person team that wrote SSNSC's mid-term rebid and developed the research programme for Phase II. Their work in Phase I positioned them well to lead the more impact-led, transdisciplinary focus emphasised by funders for Phase II. A new set of themes have been named: Blue Economy, Risk and Uncertainty, Enabling EBM, Degradation and Recovery, and Tangaroa. Three of the five theme leaders are social scientists, including one of the authors of this paper who leads the Blue Economy Theme and the Innovation Fund portfolio of eight competitively funded projects. The Tangaroa theme is now also led by a critical social scientist. Eight of the 14 core projects across all themes are led by social scientists, while all core projects include significant roles for social scientists. *Blue Economy, interpreted in sustainability transition terms, is now seen as coupled with EBM at the centre of the Challenge* (SSNSC 2020) and as a central platform for understanding interactions between the social and ecological. In the next section, we trace its travel to the heart of the Challenge.

8.6 Creating Value in a Blue Economy: Shaping the "Blue Economy Moment"

The SSNSC was rolled out as the idea of Blue Economy emerged globally as a significant discursive phenomenon (Silver et al. 2015; Winder and Le Heron 2017a). It has subsequently proliferated in institutional and academic settings, meaning different things in different places, and becoming attached to an array of political projects from coastal community development in the Global South to opening technoscientific capitalist frontiers in the EU (McKinley 2021). In the first months of SSNSC, it became clear that the Challenge would have to address the marine economy. The idea of Blue Economy gave it a potentially disruptive language to do so. *Creating Value in a Blue Economy* (CVBE) was developed by the authors of this paper as a Phase I core project. It aimed to identify the diversity of economic activities in NZ's marine spaces and investigate the connections among them—from investment to trade, regulation, and competition for resources. *CVBE asked what it would take to enact a Blue Economy that sustains or enhances the resourcefulness of environments and their communities.* It began from the hypothesis that this would require innovations in thinking, investment, production, regulation, and modes of behaviour. In these ways, CVBE created a new object of management for the increasing governmental interest in the coasts and oceans (Lewis 2018; Lewis et al. 2021b).

Our approach questioned the value of conventional representations and accounts of economy that reduce and abstract "the economy" into measures of market-based transactions and established sector categories (Lewis et al. 2021b). We rejected the reduction of economic and normative goals to growth and the language of "stakeholding." Instead, we interpreted economy in relational and holistic terms as the stewarding of resources and relations to generate long-term livelihoods for communities (Mitchell 2008) and sought to investigate "environment-economy" relationalities and the subjects, objects, and institutions involved. To do so, we made a series of departures from conventional analysis (Lewis et al. 2020, 2021b), as follows: focused attention on diversity and experimentation; treated economy as bound in co-constitutive relation to government, politics, culture, and environment; treated all actualised economies as in the making rather than prior objects to be measured, managed, and directed; adopted an "enactive research" approach, interviewing, conversing, workshopping with business, regulatory and environmental actors and community and iwi groups over a five-year timeframe; and treated each participant as an economic actor. In this way, we positioned economies as open to change and management as inherent within economy, inverting the requirement to take direction from stakeholders to a *working with economic actors*. We explored what must be achieved to foster and create value through Blue Economy. We set out to encourage transitions in environment-economy relations on an open terrain of political-investment-ecological dynamics, rather than produce tools for managerial action in predetermined fields.

Field research identified a diversity of economic activity, processes of change (*ethico-political revaluations*), and *experiments in doing economy differently*. These included policy interventions and new science funding associated with blue-tech and the development of green technologies in existing large-scale food harvesting and farming. We also found bottom-up experimentation with new practices among smaller scale circular economy seaweed and seafood initiatives, green corporate reporting, commitments to improved environmental performance via certification and promises programmes, and interest in green finance and restorative economies (Lewis et al. 2021b). These initiatives linked producer-led environmental ethics, green consumerism, value-added opportunities, policy interests in climate change, and community demands for improved community-environment outcomes. Led by indigenous ethical co-ordinates, Māori economic entities were at the forefront of many initiatives.

The research made visible a diversity of activity and relations that is normally missed by conventional treatments of economies and associated forms of management. Our approach allowed us to reorganise economic activities into new unfamiliar, overlapping, and loosely bounded categories and to offer up new relational objects for management and/or measures of success (Lewis 2018). It also confirmed that conventional categories are not only political constructs, but also perform politically, and that economies can be "made" differently. It led us to insist on a conceptualisation of Blue Economy that fractured the adversarial binaries within which established politics and management of resource use are framed—regulation-production, state-private, and management-economy. *We began to talk instead of environment-economy relations.*

The CVBE project alerted us to the disruptive possibilities of the *Blue Economy moment*, which we interpreted as a political resource for doing economy differently in Aotearoa New Zealand. The novelty, openness, and rapid proliferation of the Blue Economy category meant that it had not yet been captured by institutional frames. Instead, talk of a Blue Economy pointed to opportunities to rethink marine economy futures by leveraging the politics of access to common pool resources, concerns with climate change and environmental degradation, and gaps in scientific knowledge. The ambiguous meanings and diverse use of the term protected it from technocratic capture, made it a resource for a disruptive project of knowledge production, and enlivened its potential ecological politics. We saw the term as an opportunity to couple the growing recognition of the global significance of the oceans as a site of economy critical to humanity's needs with a response to the environmental degradation of the oceans and the politics of community development. This possibility is what we term "the Blue Economy moment."

The Blue Economy moment opens up possibilities to create conditions for doing differently, affirming the rationale for disruptive research in a language that reassures government and industry stakeholders. As Phase I became Phase II and CVBE transformed into the Blue Economy Theme, we began to narrate Blue Economy to stakeholders as *less an object and more an aspiration linked to just transitions*. To address the definitional imperative of stakeholders, we identified *four core propositions common to diverse interpretations of Blue Economy*: the oceans are central to humanity's needs and a source of economic opportunity; the oceans are in crisis; addressing needs, realising opportunities, and confronting crisis require new socio-ecological knowledge and new regulation; and just transitions are necessary and possible but require active interventions. In these terms, we worked with colleagues to develop four core projects for Phase II and redefine the notion of innovation for a Blue Economy in funding a new round of Innovation Fund proposals. The core projects funded seek to indigenise Blue Economy and create the conditions for each of the elements of restorative economy, marine ecotourism, and a new seaweed sector premised on EBM (https://www.sustainableseaschallenge.co.nz/our-research/blue-economy/). The Innovation Fund projects redefine innovation and its possibilities. They include innovation for community-based, diverse, and mātauranga Māori-driven economic activities. While three projects explore the utilisation of biotech and ICT to develop products and processes, the others are centred on ecological co-benefits (utilising pest species, collective ownership of rents, and reducing the environmental harms of current practices). In short, the Theme has been built as a platform for delivering a new way of thinking and doing economy in the oceans.

8.7 A Knowledge Infrastructure for a Transformative Blue Economy?

The sections above document the assemblage of a marine social science community in relation to opportunities offered by SSNSC and the Blue Economy moment. In short, social scientists *seized unforeseen opportunities to experiment in the novel knowledge spaces created by a new science investment stream that opened-up disciplinary and institutional boundary lands.* In CVBE, the authors explored the generative possibilities of treating the economy as a situated assemblage of activities, actors, institutions, and relations grounded in social life rather than located in a radically separated sphere. The work elevated questions of economy and helped lead the broader *social scientisation of SSNSC.* It yielded a disruptive knowledge that rescripted economy, science, and management as co-constitutive rather than detached domains.

The CVBE project leveraged the rare intellectual freedoms and institutional support provided by SSNSC and the Blue Economy moment to disrupt business as usual through a philosophy of *knowing economy differently to do economy differently.* Interpreting *Blue Economy as an aspiration, challenge, and opportunity* opened up a politics of transitioning and allowed SSNSC to prioritise aspirational and transformational projects for funding in Phase II rather than projects that would support existing management of predefined objects. We devised research questions and funded projects that demanded transformative concepts and methodologies, promised to disrupt relations between knowledge and business as usual, and redirected imperatives to conduct impact-focused research towards *doing differently.* In this sense, social science in SSNSC aligned a set of contingencies and opportunities to assemble a robust configuration of people, artefacts, and institutions into a *knowledge infrastructure for transitioning* to more sustainable ocean futures in NZ. Knowledge infrastructures organise the interplay between expertise, its exercise, and the discursive formations, networks, funding structures, and political connections that support social life (Edwards 2017). They frame social practice, shape political participation, and configure the decision-making that sustains it. The SSNSC is in these terms an infrastructure for assembling knowledge practices, content, and performative effects for knowing and managing the resourcefulness of environment-economy relations.

We contend that social science, Blue Economy research and the knowledge base being built around the management of environment-economy relations have become crucial to this infrastructure. In material terms, SSNSC as a knowledge infrastructure is composed of research projects, funding flows, and networks of researchers, which are held together by an overarching objective, interconnected themes, commitments to transdisciplinary and co-produced knowledge, common research practices, periodic moments of assembly, a carefully curated website, and an organisational structure of theme leaders, directorate, governance groups, stakeholder panels, and a secretariat. It is, however, neither closed nor stable. Like other infrastructures SSNSC is lively, uncertain, and in the making—a brittle and incomplete infrastructure for transitioning. *It is an achievement* but one that was difficult to achieve and falls short of ideal in practice.

How has all of this been achieved? The short answer is that nearly two decades of boundary-crossing, enactive research in multiple settings taught us that we should seize as they surface, work with contingency, enlist allies, and build networks. In our work with SSNSC we have built close relationships with ecologists in multiple universities and CRIs, and with social scientists who do different work to ours. Our experiences point to the potentiality of experimentation but also to the limits that brute political economy can impose on possibility. We have not always succeeded. We failed, for example, in our CoRE bid. We continue to struggle against the intransigencies and politics of established interests and knowledge categories within SSNSC and its broader community of stakeholders. The Blue Economy suite of project concepts, for example, was rejected in early 2019, forcing us into a further year of research and a redevelopment of projects into more acceptable forms.

However, we cannot point decisively to evidence that we have altered key decision-making frameworks or specific decisions. Ecologically destructive activities continue, and economic rents are still being extracted unjustly from communities and their commons. Other science funding is still being directed at supporting growth agendas, and colleagues still at times represent Blue Economy in the more conventional terms of growth and industry sectors. It is unclear what will happen from mid-2024 when SSNSC funding ends.

8.8 Conclusion: Building a Knowledge Infrastructure through Social Science

The chapter tells two interwoven stories: one of building a platform for transitioning to new economies, and one of the potentiality of communities of critical social science. Woven together they describe a political project of knowledge production that has sought to enact transitions by shaping new knowledge assemblages in institutional and disciplinary border spaces. As key characters in the story, we did not set out to do this explicitly. Rather we grasped opportunities to experiment and participate in unfamiliar spaces and worked to centre social science, transcend its presumed roles, and reconfigure its relationships with science, policy, and social change. We assumed intellectual and institutional leadership and sought to replace knowledge categories that close-down thinking and serve established power with new configurations of power-knowledge. To do this we fostered new relationships with knowledge makers, brokers, and users and drew on critical theory to experiment with more flexible, practice-based conceptual framings. These allowed us to navigate the "borderlands" of SSNSC and rework its "growth within limits" agenda into an ethos of *resourceful environment-economy relations for just transitions* founded on new practices and relations of knowledge production.

Building the community took time and alignments with others. It also took the intellectual, institutional, and political opportunities provided by the Blue Economy moment to gain otherwise unlikely access to decision-making and float new and unfamiliar concepts into diverse spaces of influence. Our narrative shows, however, that in the right conditions social scientists can transcend institutional constraints to generate knowledge interventions that are simultaneously conceptually adventurous, practice-oriented, and politically animated. They can create transformative economic knowledge, build a novel knowledge infrastructure, and initiate transitioning pressures that influence investment trajectories. How, though, should we judge the foundations of an argument based on our own analysis of our own work?

As post-structuralists we are confident that by reshaping knowledge infrastructures and engaging enactively with actors beyond the academy, we cannot help but change worlds (and economies) that are always in the making (Law and Urry 2004). More prosaically, the networks, knowledge bases, and capabilities built through SSNSC exist outside of us and are having performative impacts on landscapes of knowing and doing oceans economy. They have shaped the direction of an $80 million research programme, directing upwards of $20 million into grounded social science research, including in unlikely domains such as reviving traditional forms of aquaculture, community economy experiments in aquaculture, and restorative economy. Social science from SSNSC has been communicated directly into legislative reform processes, regional development debates, and public and private investment trajectories. SSNSC can claim some credit for stimulating the 2020 formation of a new national Ministry for Oceans and Fisheries and for invigorating efforts to implement participatory governance initiatives in multiple settings and at multiple scales.[1] These initiatives reframe the management and governance of environment-economy relations and dislodge the grip on the oceans held by growth-focused interests. Finally, social scientists have circulated Blue Economy and other narratives into public spaces via multiple media—stakeholder reports, public webinars, public-facing values-ends-means workshops, educational videos, and an SSNSC website richly adorned with objects and insights from social, socio-economic, and socio-ecological research (Greenaway et al., 2019).

Beyond the specifics of our case, we can offer insights into the potential for knowledge making in emergent borderland knowledge spaces anywhere:

- 'Think/do/be' proactive and prominent—identify and seize opportunities in academic, policy, institutional, and material world settings, and practice a micro-politics of knowledge making that translates knowledge into impact
- Adopt enactive research methodologies and strategies that target the inherently impactful nature of engagement, embrace the co-production of knowledge, and create opportunities to influence political and economic actors (every meeting/interview/workshop is an intervention opportunity)
- Graft hard in institutional spaces and work to keep options open, expand networks, align with others, and foster capacities/capabilities

- Operate consciously across epistemological, disciplinary, and institutional boundaries to supplant existing knowledge architectures with new cross-boundary infrastructures
- Have the confidence to experiment, reset conceptual framings, seize the intellectual agenda, and eschew easy existing explanatory metaphors and categories

These lessons are neither easily drawn nor pre-packaged for use elsewhere. They emerge from 20 years of enactive research and network and capability building grounded in poststructuralist dispositions: a situated politics of possibilities, a methodology of knowing by doing, and an onto-epistemology of practice in worlds that are in the making (Lewis et al. 2016). If we are serious about building a marine social science for transitions, we must take the lead from deeply engaged initiatives such as SSNSC, Heidkamp and Morrissey's coastal transitions programme (2018), and Voyer's Australian work (Voyer et al. 2021), and encourage others to create new architectures and infrastructures of power-knowledge in other places.

The SSNSC has animated a community of social scientists. When it began, there was effectively no dedicated marine social science in Aotearoa New Zealand. There are now up to 25 marine social scientists at major research organisations and at least an equal number of ecologists, indigenous scholars, and consultants producing marine social science in different settings—all connected into and through SSNSC. Multiple socio-ecological PhD and postdoctoral projects are underway. The community has a prominent presence in key international academic debates. We agree wholeheartedly with McKinley's call to proliferate such communities and "better understand" the complexity of actors, multiplicity of values, and diversity of relationships among people and the sea involved in blue economies. Yet, we challenge her to go further along several axes: from human actors to more than human actors and actions, disciplinary to inter- or transdisciplinary knowledge, categorical to practice-based knowing, and from talking of a "need for," marine social science to doing it and demonstrating its value. The "need" at hand is to create a new knowledge infrastructure that escapes prior categories such as sectors and disciplinary fields as well as the declarative thought, action, and relations they produce—both in understanding Blue Economy and in *practising social science to disrupt power-knowledge relations and do Blue Economy differently.*

Finally, we ask where we would be if these two human geographers and their collaborators hadn't pushed themselves into these border spaces and used them to experiment? We can claim confidently that SSNSC would have been less experimental, theoretically innovative, inter-disciplinary, and attuned to relationality; and more constrained by scientific rationalities, established language and categories of knowledge, growth within limits thinking, and a self-destructive contest between environment/management and economy/business. That is significant.

Acknowledgements

This research was funded by the New Zealand Ministry for Business Innovation and Employment (C01X1515) through the Sustainable Seas National Science Challenge's Valuable Seas project 2.2.1 "Creating Value from a Blue Economy." We thank all our colleagues within the Challenge for their endeavour.

Note

1 https://www.mpi.govt.nz/dmsdocument/45550-Revitalising-the-Gulf-Government-action-on-the-Sea-Change-Plan

References

Bargh, M. (2014). A Blue Economy for Aotearoa New Zealand? *Environment, Development and Sustainability, 16*(3), 459–470.

Bennett, N. J. (2019). Marine social science for the peopled seas. *Coastal Management, 47*(2), 244–252.

Carolan, M. (2013). The wild side of agro-food studies: On co-experimentation, politics, change, and hope. *Sociologia Ruralis, 53*(4), 413–431.

Davies, K., Fisher, K., Foley, M., Greenaway, A., Hewitt, J., Le Heron, R., Mikaere, H., Ratana, K., Spiers, R., & Lundquist, C. (2018). Navigating collaborative networks and cumulative effects for Sustainable Seas. *Environmental Science & Policy, 83*, 22–32.

Edwards, P. N. (2017). Knowledge infrastructures for the Anthropocene. *The Anthropocene Review, 4*(1), 34–43.

Hamilton-Hart, N., & Stringer, C. (2016). Upgrading and exploitation in the fishing industry: Contributions of value chain analysis. *Marine Policy, 63*, 166–171.

Greenaway, A., & the CVBE Team. (2019). *Generating a 'Blue Economy' through ecosystem-based management (EBM)*. Sustainable Seas: Wellington. https://www.sustainableseaschallenge.co.nz/tools-and-resources/generating-a-blue-economy-through-ebm/

Heidkamp, C.P., & Morrissey, J. (Eds.). (2018). *Towards coastal resilience and sustainability*. Oxford, UK: Routledge.

Law, J., & Urry, J. (2004). Enacting the social. *Economy and Society, 33*(3), 390–410.

Le Heron, R., & Lewis, N. (2011). New value from asking 'Is geography what geographers do?' *Geoforum, 42*(1), 1–5.

Le Heron, R., Rees, E., Massey, E., Bruges, M., & Thrush, S. (2008). Improving fisheries management in New Zealand: Developing dialogue between fisheries science and management (FSM) and ecosystem science and management (ESM). *Geoforum, 39*(1), 48–61.

Le Heron, R., Lewis, N., Fisher, K., Thrush, S., Lundquist, C., Hewitt, J., & Ellis, J. (2016). Non-sectarian scenario experiments in socio-ecological knowledge building for multi-use marine environments: Insights from New Zealand's Marine Futures project. *Marine Policy, 67*, 10–21. https://doi.org/10.1016/j.marpol.2016.01.022

Le Heron, E., Le Heron, R., Blackett, P., Davies, K., Logie, J., Allen, W., Greenaway, A., & Glavovic, B. (2019). It's not a recipe… but there are ingredients. Navigating negotiated changes through participatory processes in marine spaces. *Planning Quarterly, 213*, 32–37.

Le Heron, E., Le Heron, R., Taylor, L., Lundquist, C., & Greenaway, A. (2020). Remaking ocean governance in Aotearoa New Zealand through boundary-crossing narratives about ecosystem-based management. *Marine Policy, 122*. https://doi.org/10.1016/j.marpol.2020.104222

Le Heron, E., Allen, W., Le Heron, R., Logie, M., Glavovic, B., Greenaway, A., ... & Blackett, P. (2021). What does success look like? An indicative rubric to assess and guide the performance of marine participatory processes. *Ecology and Society*, 26(1). https://doi.org/10.5751/ES-12211-260129

Lewis, N. (2018). Cultivating diverse values by rethinking Blue Economy in New Zealand. In J. Morrisey and P. Heidkamp (Eds.), *Coastal transitions: Towards sustainability and resilience in the coastal zone*. Oxford, UK: Routledge.

Lewis, N., & Le Heron, R. (2019). Poststructural political economy. In A. Kobayashi (Ed.), *International encyclopedia of human geography*, 365–373.

Lewis, N., Le Heron, R., Carolan, M., Campbell, H., & Marsden, T. (2016). Assembling generative approaches in agri-food research. In R. Le Heron, H. Campbell, N. Lewis, M. Carolan (Eds.), *Biological economies: Experimentation and the politics of agri-food frontiers*. Oxford, UK: Routledge, 11–29.

Lewis, N., Le Heron, R., Hikuroa, D., & Le Heron, E. (2020). *Making new Blue Economy in Kaikōura: A participatory process approach*. Wellington: Sustainable Seas National Science Challenge, 1–21.

Lewis, N., Le Heron, R., Hikuroa, D., Le Heron, E. (2021a). Remaking regional development. Blue Economy rent platforms in Kaikoura, Aotearoa New Zealand. *Regional Studies* (Submitted).

Lewis, N., & the Blue Economy research team (2021b). *Creating value in a Blue Economy*. Wellington: Sustainable Seas National Science Challenge, 1–140.

McCormack, F. (2018). Māori saltwater commons. *Commoning Ethnography*, *1*(1), 9–31.

McKinley, E. (2023). A Blue Economy for who? Linking marine social sciences with Blue Economy discourse. In C. P. Heidkamp, J. Morrissey, C. Germond-Duret (Eds.), *Blue Economy: People and regions in transitions*. Oxford, UK: Routledge.

McKinley, E., Acott, T., & Yates, K. (2020). Marine social sciences: Looking towards a sustainable future. *Environmental Science and Policy*, *108*, 85–92.

Mitchell, T. (2008). Rethinking economy. *Geoforum*, *39*(3), 1116–1121.

Peart, R. (2018). *Voices from the Sea: Managing NZ's fisheries*. Auckland: Environmental Defence Society.

Reid, J., Rout, M., & Mika, J. (2019). *Mapping the Māori marine economy*. Wellington: Sustainable Seas National Science Challenge.

Roelvink, G., Martin, K. S., & Gibson-Graham, J. K. (Eds.). (2015). *Making other worlds possible: Performing diverse economies*. Minneapolis, MN: University of Minnesota Press.

Rout, M., Awatere, S., Mika, J. P., Reid, J., & Roskruge, M. (2021). A Māori approach to environmental economics: Te ao tūroa, te ao hurihuri, te ao mārama—The old world, a changing world, a world of light. In *Oxford Research Encyclopedia of Environmental Science*. Oxford: Oxford University Press. Oxford.

Ruru, J. (2018). Listening to Papatūānuku: A call to reform water law. *Journal of the Royal Society of New Zealand*, *48*(2–3), 215–224.

Silver, J., Gray, N., Campbell, L., Fairbanks, L., & Gruby, R. (2015). Blue Economy and competing discourses in international oceans governance. *The Journal of Environment & Development*, *24*(2), 135–160.

Sustainable Seas National Science Challenge (SSNSC) (2020). Creating value from a blue economy. The University of Auckland.

Šunde, C., Sinner, J., Tadaki, M., Stephenson, J., Glavovic, B., Awatere, S., Giorgetti, A., Lewis, N., Young, A., & Chan, K. (2018). Valuation as destruction? The social effects of valuation processes in contested marine spaces. *Marine Policy*, *97*, 170–178.

Thrush, S., Lewis, N., Le Heron, R., Fisher, K., Lundquist, C., & Hewitt, J. (2016). Addressing surprise and uncertain futures in marine science, marine governance, and society. *Ecology and Society*, *21*(2), 44 [online]. http://www.ecologyandsociety.org/vol21/iss2/art44/

Ulrich, S., Thrush, S., Hewitt, J., Jorgensen E. (2018) The Earth Summit 25 Years on: Why is biodiversity continuing to decline? Resource Management Law Association of New Zealand.

Voyer, M., Moyle, C., Kuster, C., Lewis, A., Lal, K., & Quirk, G. (2021). Achieving comprehensive integrated ocean management requires normative, applied, and empirical integration. *One Earth*, *4*(7), 1016–1025.

Winder, G., & Le Heron, R. (2017a). Assembling a Blue Economy moment? Geographic engagement with globalising biological-economic relations in multi-use ocean environments. *Dialogues in Human Geography*, 7(1), 3–26.

Winder, G., & Le Heron, R. (2017b). Further assembly work: A mountains to seas Blue Economy imaginary. *Dialogues in Human Geography*, 7(1), 50–55.

9

THE BLUE ECONOMY, CLIMATE, TOURISM, AND SOCIAL INJUSTICE IN BARBUDA

José A. Torres

9.1 Introduction

The 2017 hurricane season brought storms that accounted for unusually high levels of destruction. Hurricane Maria brought attention to the US territory of Puerto Rico for the magnitude of its destruction that included an estimated 472,000 destroyed or severely damaged homes (Habitat for Humanity, 2021) in an island with a present-day population of around 3.2 million (Central Intelligence Agency (CIA), 2021) that does not include the estimated 130,000 who left the island after the hurricane (CNN, 2018). Hurricane Irma resulted in devastating losses that received almost no attention in spite of the fact that it accounted for the destruction of around 95% of all built structures and for wiping out the island's vegetation cover and forced the evacuation of the entire population in the island of Barbuda, part of the Caribbean nation of Antigua and Barbuda (NPR, 2020; The Conversation, 2019; IAI, 2018).

Plans for reconstruction in the island nation have resulted in a controversial plan to radically alter land ownership laws in Barbuda, from communal land ownership to privatisation of ownership of the land that is to be made available not only to Barbudans but also to Antiguans and individuals from all around the world. This book chapter discusses this attempt at changing land ownership in Barbuda as an instance of social injustice that runs counter to what should be happening in region whose economic development plans are guided by environmentally and socially sustainable Blue Economy practices.

9.2 Antigua and Barbuda

The nation of Antigua and Barbuda is located in the Leeward Islands region of the Lesser Antilles (Figure 9.1), and it also includes the small uninhabited island of Redonda. The island of Antigua has a population of approximately 95,000

DOI: 10.4324/9781003280248-12

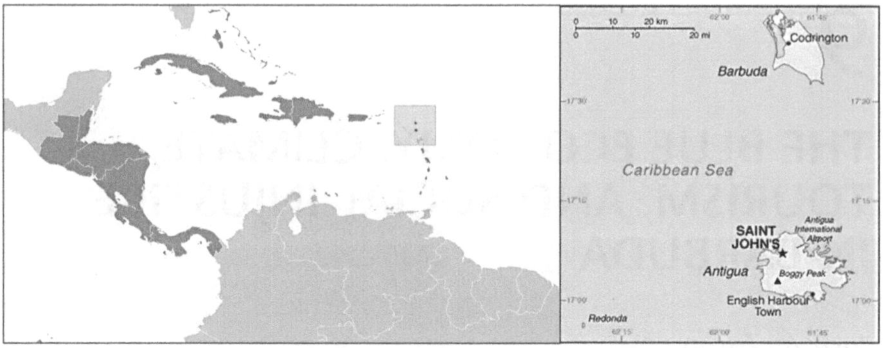

FIGURE 9.1 Antigua and Barbuda (CIA, 2021)

inhabitants, while the pre-Irma population number for Barbuda was close to 3000. In terms of land area, Antigua covers 108.1 square miles, and Barbuda covers 62.1 square miles (CIA, 2021). A familiarity with the region and an examination of satellite and topographic maps shows that like most islands in the Caribbean, Antigua is of volcanic origin and has a mountainous terrain. Barbuda, on the other hand, is of coral build-up origin, and the terrain is flat and barely above sea level. Only two other islands in the arch of Caribbean islands that includes the Greater Antilles, and the Leeward and Windward Islands of the Lesser Antilles are also of coral origin and flat. These islands are Anguilla, and Anegada in the British Virgin Islands. As a matter of fact, *anegada* is a Spanish word that means "submerged" or "flooded." Tourism accounts for 60% of the Antigua and Barbuda's GDP, and tourism and services employ 82% of the nation's labour force. Construction, light industry, and small amounts of agricultural activity also contribute to the nation's economy (CIA, 2021). The government of Antigua and Barbuda does not provide separate tourism figures for each island (Research and Statistics Department/Ministry of Tourism, Foreign Affairs and Immigration of Antigua and Barbuda, 2017), but Antigua is a popular cruise ship stop and also a popular beach resort destination, while Barbuda tends to be considered an off-the-beaten track type of destination. Being an island nation, Antigua and Barbuda is the type of location for which for which Blue Economy policies for sustainable development have been developed.

9.3 The Blue Economy, the Tourism Industry, and Social Justice Theory

What is today known as the Blue Economy originated as Goal #14, aimed at Life Underwater, out of 17 Sustainable Development Goals that were unveiled by a United Nations conference in 2012 in Rio de Janeiro. First known as Blue Growth and now known as the Blue Economy, this movement originally paralleled Green Initiatives in terms of its concern with environmental issues, but aimed specifically at oceans, seas, major lakes, coastal areas, and river estuaries. More recently, the parallels between Green Initiatives and the Blue Economy now include socioeconomic factors and impacts in the sustainability goals and measures that are part of Blue Economy initiatives (Garland et al., 2019; Graziano et al., 2019).

The tourism industry that develops in coastal and island locations attracts large numbers of people seeking the following: the sea, sun, and sand of tropical and warm subtropical locations; the opportunity to admire visually appealing landscapes that combine the land and the sea; relaxation possibilities; and/or the recreational and entertainment possibilities that coastal and island locations can offer. The fact that these locations have a high level of appeal makes the locations vulnerable to over-tourism and the effects that over-tourism may have on coastal and marine environments. The tourism industry in coastal and island locations relies heavily on the appeal and the good visual and environmental health of beaches, shores, and adjacent bodies of water if the industry is to be a long-term option as an economic activity instead of being a self-destructing, short-lived option.

Another important effect of a place having high levels of tourist activity in coastal and island locations is the effects that visits by outsiders have on local populations. Based on factors of location and seasonality, the industry provides low-wage employment, even among employees of high-end resorts and hotels, and in locations where weather and climate provide seasonal fluctuations in the number of visitors the low-wage employment tends to be seasonal as well, denying employees a guaranteed year-round income, as low as it may be.

The loss of coastlines to tourism sector property owners may in return bring only limited-to-minimal economic benefits to the popular coastal and island locations if there are few linkages between hotel and resort owners and their places of operation. In the worst-case scenario, linkages are limited to hotel and resort owners who only contribute to the local economy in terms of payroll, utility payments, and taxes if they are not able to obtain tax credits or exemptions when setting up their facilities. Other linkages are determined by how much of the local production or retail is used in the daily operations of the facilities, such as food, furniture, linens, cleaning products, and other supplies and services. Money that is not linked to the local economies is money that is sent elsewhere in the form of imports or repatriation of profits.

In locations where the population is subject to unusually high levels of unemployment, low wages, and/or limited prospects for earning an income, the presence of visitors who are or appear to be affluent, at least compared to the local quality of life indicators, may create another socioeconomic issue that links to tourist activity, that of crime that may range from crime of opportunity or petty crime to much more violent crime that targets visitors.

Coastlines that are turned into hotels and resorts turn into land and sea that suddenly become off-limits to local populations who would otherwise make use of land and marine resources for their own economic sustenance or for recreation. Locals lose access to those resources while hotel owners, whether they are independent or corporate, and most likely foreigners, become the new masters of useful land and the adjacent sea that may be in short supply in smaller and/or isolated locations. The loss of access and rights to land and sea by local communities brings the issue of social justice to this discussion. Though the topic of social justice can be perceived in simple terms of what is or appears to be right and what is wrong, the study of social justice has been defined by a theoretical framework that brings uniformity and objectivity to the discussion of the topic within any type of context.

Rawls' political philosophy of justice understands the concept as a distribution of rights and duties in a society, and a mutually acceptable understanding of justice lies in the determination of what is the proper distribution of benefits and burdens of social cooperation within any society (Sabbagh & Smith, 2018; Rawls, 1999). A different type of understanding of the concept of justice is based on the psychological perceptions of the concepts of equity (proportional representation and influence in a group or society) and equality (same access to opportunity), or lack thereof (Sabbagh & Smith, 2018) and justice. The psychological perceptions of equity and equality also imply an awareness of the equitable distribution of burdens as well as the benefits.

The concept of environmental justice establishes that there should be an even distribution of environmental costs or environmental effects within societies (Sabbagh & Smith, 2018; Nweke & Lee, 2014), and the same can be said about climate justice. Forcing groups to take a disproportionate amount of burden in order to react to after the fact, or prevent, the effects of climate change, climate and weather activity, and natural and human-made environmental issues based on location, socio-economic status, or ethnic, racial, and/or cultural considerations is a clear example of what is considered environmental or climate injustice. Environmental and climate justice has become an integral part of any measure of sustainable development and sustainable practices such as those set forth by Blue Economy policies, and which are also those set forth by the Inter-American Development Bank (IDB) in its platform to guide sustainable development in islands in the Caribbean region.

9.4 The Inter-American Development Bank & The Sustainable Islands Platform

The IDB developed its Sustainable Islands Platform (SIsP) "to promote sustainability among island territories through innovative development pathways" (IDB, 2021), focusing on economic, environmental, and social sustainability. The SIsP is built on three conceptual pillars, the Blue Economy, the Circular Economy, and Climate Resilience (IDB, 2021). The Blue Economy is discussed in the previous section, and it is the Pillar that is presented first in the Description of the Platform.

The second Pillar, the Circular Economy, is a Pillar that is not exclusive to sustainable goals in either island territories or the Western Hemisphere but is an integral part of any sustainability program. In the Circular Economy, the aim is to eliminate as much as possible single-use materials such as plastics used for wrapping and for placing products on the way to the cash register in a grocery store or any other materials designed for single- or limited-use purposes and not biodegradable. The Circular Economy favours aggressive recycling activities on order not only to extend the life of used materials but also to minimise the amount of waste that is generated. It also encourages the maximisation of the use of renewable energy sources and the elimination of toxic materials and waste that harm the biosphere (IDB, 2021).

The Third Pillar, Climate Resilience, "is the capacity to prepare for, predict, and respond to natural disasters, hazardous events, trends, or disturbances related to the climate" (IDB, 2021). This Pillar borrows from the US Climate Resilience Toolkit, whose "steps to resilience include exploring hazards, assessing vulnerability and risks, investigating options, and taking action." A program developed within the Climate Resilience Pillar is the Climate Smart Islands Program, through which islands earn the Climate Smart Island designation by establishing "climate resilient measures such as energy, water, waste treatment, tourism, fisheries, transport and infrastructure, among others" (IDB, 2021), incorporating sustainable practices in all aspects of economic and daily life for the islands.

The Sustainable Islands Platform and the Climate Smart Islands Program present the islands with goals and policies that have a specific aim, establishing sustainable practices, but these goals may clash with the reality of the islands' different resources and capabilities to adopt those polices and reach those goals. Another issue to contend with is the effect that actual climate events with unexpected levels of intensity and destructiveness may have on islands, such as the aftermath of hurricane Irma's passing through the island of Barbuda in 2017, which included not only the actual catastrophic levels of destruction but also the government's controversial approach to rebuilding structures, the economy, and life on the island.

9.5 Recovery and Land Ownership after Hurricane Irma

While the Caribbean region is being subject to aggressive and well-detailed plans for sustainable approaches to environmental stewardship, and socio-economic development with social justice included, weather events can illustrate how difficult it is to implement the procedures that would lead to widely accepted and implemented sustainable practices. The passing of hurricane Irma through Caribbean islands that included Barbuda in 2017 resulted in the island's vegetation cover being wiped out, the destruction of approximately 95% of all built structures, and the temporary mandatory evacuation of the entire island's population to the island of Antigua (NPR, 2020; The Conversation, 2019; IAI, 2018). Since then, plans made and action taken to restore Barbuda's destroyed built structures have been slow, with some accusations made as to this slowness being deliberate. The most radical and contentious solution to funding the restoration of the island's structures has been to change the land ownership laws that have been in place since slaves were emancipated in 1834 (NPR, 2020; The Conversation, 2019; IAI, 2018). When slaves were emancipated, it was determined that land in Barbuda would be communally owned. "Barbudans do not own the land, but lease it to build their homes, to farm, and for economic activity" (BBC News, 2019). Persuaded by groups that have been labelled as "investors" and "disaster capitalists," the country's government, based in the island of Antigua where about 95% of the country's population is located, decided that the best way to fund the rebuilding of Barbuda would be to eliminate its communal land ownership laws and to privatise land ownership, with land values to be determined by market forces as the land is being sold and bought to individuals and groups from around the world.

Arguments made in favour of privatising land ownership included the opportunity for Barbudans to, for the first time in almost 200 years, hold land ownership titles (NPR, 2020; The Conversation, 2019; IAI, 2018). Those titles can then become an asset to Barbudans, which they can either hold on to in order to accrue more value or use as collateral for loans which would give individuals money to repair their homes and businesses in any way and form that they see fit. And if they choose to sell the land, then they will be receiving the kind of income that they most likely had never seen before. This option has some appeal, especially as efforts to rebuild Barbuda have been slow and inefficient, deliberately so in order to make privatisation more palatable (NPR, 2020; The Conversation, 2019; IAI, 2018) according to sceptics and opponents of privatisation.

Those who advocate against the privatisation of land in Barbuda argue that it is at best a short-sighted solution and at worst a policy that would end up making Barbudans transient renters in what used to be their land (NPR, 2020; The Conversation, 2019; IAI, 2018). The suspicion is that land titles would not last long in Barbudan hands. Those who would use their land titles as collateral for loans would lose their land if they found themselves defaulting on their loans. Privatised land, especially land that became privatised as a result of a small country's government being pressured by multinationals and the investor class to make coveted tropical island land available to them, is subject to speculation, being bought for the sole purpose of re-selling it instead of developing it, to be developed to meet the wishes of non-Barbudan individuals who may or may not have a year-round presence in the island, or being developed into tourist hotels and resorts that take land and water use away from the locals.

Opponents of privatisation point out that the possible uses for privatised land listed above would have an effect similar to gentrification. It would not be long before land titled changed hands enough times to make those titles' prices increase so much that they would become out of reach of any Barbudan's budget. Barbuda would then become littered with overpriced vacation homes that, just like the land itself, would also be out of any Barbudan's financial reach. Non-Barbudans would own most of the island and its properties, with Barbudans having no option but to be concentrated in Codrington, the island's town, being renters if there were any pressure to sell their homes as well. And as an island modified to fit the needs and whims of the tourist and vacation home industries, Barbudans would have little occupational options besides being low-wage tourist industry employees or being the low-paid cleaning and service employees for vacation homes.

In 2018 (BBC News, 2019), the legislature of Antigua voted in favour of allowing the privatisation of the land in Barbuda. The only reason the new land ownership law is not in effect is because Barbudans sued to stop the implementation of the new law. As of early 2021, land in Barbuda has yet to be privatised. Though the law has not taken effect yet, this kind of legislation presents a case of social injustice taking place at the same time that Caribbean islands are being provided with planning tools that facilitate sustainable socio-economic development practices and policies. This instance of social injustice is not based on race, ethnicity, or cultural features such as language or religion. Barbudans became a disadvantaged group because of their relative location, the smaller or two inhabited islands, and because their island's population amounts to roughly 3% of the entire country's total. Barbudans became a minority group due to unusual circumstances that usually would not be the justification for having minority status forced upon them.

From the political philosophy and the environmental justice standpoints of social injustice, the law that allows land privatisation leaves Barbudans out of the decision-making and rebuilding processes in their own island, and it guarantees that the handling of revenue for taxes and fees resulting from the selling and reselling of the Barbudan land will be controlled, from collection to redistribution, by the Antiguan side of the island country. This approach by the government also runs counter to psychological and environmental perceptions of social justice guaranteeing an equitable distribution of burdens and benefits in the reconstruction of Barbuda after a catastrophic weather event. The outcome of this instance of social injustice needs to be monitored because the occurrence and the magnitude of catastrophic climatic events keeps increasing in the Caribbean. The outcome of the situation in Barbuda may influence the present and future to redevelop areas of islands that are devastated by these climate events in the age of Blue Economy and social justice considerations in a region that is seeing uneven approaches to sustainable socioeconomic initiatives such as the ones set forth by the Blue Economy.

9.6 Progress and Concerns in the Caribbean

There are instances of the Blue Economy serving as a real blueprint for socioeconomic development in the Caribbean. A significant instance is taking place the island of Bonaire, located close to Aruba and Curaçao, and part of what is used to be the Netherlands Antilles that now hold the status of municipality within the Netherlands (CIA, 2021). Bonaire has declared itself a Blue Destination, the world's first and so far, only tourism destination to label itself as Blue based on Blue Economy guidelines. The Blue Destination designation is displayed and explained thoroughly in the island's tourism websites (Blue Destination Bonaire, 2021; Tourism Bonaire, 2021), and it makes evident what the island's policies and priorities are as a tourist destination, and it also sets the tone for not only what visitors should expect to encounter in a Blue Destination but also for what the role of a visitor in a Blue Destination needs to be. The Blue Destination designation is more than just a label. It sets the framework for what the tourism experience will be for visitors and locals alike.

On the other hand, Puerto Rico presents a case of environmental injustice that indicates that the case in Barbuda is not an isolated case. In order to help rebuild the island's economy after hurricane Maria, which swept through the Caribbean shortly after hurricane Irma in 2017, the commonwealth's government passed legislation intended to promote investment from non-residents and to attract wealthy non-residents with the promise that they would not pay any taxes on money brought into and/or invested in the island. By early 2021 the results of the policies can be noticed. In the historic core of San Juan, founded in 1521 and continuously inhabited since then (Nuestro Puerto Rico, 2021), at least 24.3% of all residential space is now owned and controlled by the short-term tourist rental sector. In the island of Culebra, east of the main island, at least 27.4% of all residential space is also controlled by the short-term tourist rental sector (Latinx Today, 2019). Puerto Rican residents are being displaced from rental units and denied housing options so that property owners, including recently arrived investors, can generate higher monthly rental incomes. Additionally, coastal lands that had been set aside to protect their natural and environmental assets are now lands where the land local and non-Puerto Rican developers, and tourism and hospitality corporations, none of which are headquartered in Puerto Rico, are now able to modify and build on as they please, in spite of protests from organised opponents of the destruction of what had been set aside as protected land (Eyboricua, 2020).

9.7 Conclusion

The Caribbean region has not escaped the effects of climate change that is occurring in the form of tropical storm activity that is increasing both in frequency and intensity. Though there are programs that incorporate Blue Economy guidelines into development plans that include environmental and socioeconomic sustainability, political and economic realities in the islands leave them vulnerable to outside interests when it is time to rebuild them after they are damaged by tropical storms. Disaster capitalists who come in the form of investors, land developers, and/or hotel and resort operators, all motivated by the prospect of profits derived from the tourism and hospitality industry, know how to take advantage of Caribbean governments and the region's reliance on the tourism and hospitality industry to develop their economies when they identify new business opportunities that can be promoted as conveniently-timed injections of money that will help in post-storm redevelopment efforts. Whether it is out of necessity or even desperation, or because disaster capitalists are experienced when it comes to communicating with the governments of sovereign countries and of territories, islands in the Caribbean are likely to continue to be pressured to ignore environmental and socioeconomic concerns when faced with offers to invest in local economies post-storms. It must be noted that, as of January 2021, there haven't been reports about how tourism industry investors, land developers, and facilities operators have changed strategies on new opportunities since the COVID-19 pandemic drastically reduced the amounts of in-country and international tourist activity. Once this information becomes available, attention should be given to how changes in the tourism industry during and post-pandemic affect the issues of social injustice in Barbuda as well as other places in the Caribbean such as Puerto Rico, and if the pandemic results in the strengthening rather than the weakening of the role that the Blue Destination designation plays in the tourism industry in Bonaire.

References

BBC News (2019), Why I don't want to own the land my business is built on, https://www.bbc.com/news/stories-49210150#:~:text=Barbudans%20do%20not%20own%2C%20but,of%20history%2C%22%20says%20Frank, last accessed January 25, 2021.

Blue Destination Bonaire (2021), Bonaire, the first Blue Destination of the World, https://bluedestination.com/, last accessed January 25, 2021.

Central Intelligence Agency (CIA), https://www.cia.gov/the-world-factbook/, last accessed January 25, 2021.

CNN (2018), 130,000 left Puerto Rico after Hurricane Maria, Census Bureau says, https://www.cnn.com/2018/12/19/health/sutter-puerto-rico-census-update/index.html, last accessed January 25, 2021.

Eyboricua (2020), Polémico reglamento permite comercios en reservas naturales, https://eyboricua.com/noticias/puerto-rico/polemico-reglamento-permite-comercios-en-reservas-naturales/?fbclid=IwAR2iZac7wo3pr7ta4m1oUtcAXVWUrL0N-HdBczeGkbjJ4priHBXZjmS7UWdc, last accessed January 25, 2021.

Garland, M., Axon, S., Graziano, M., Morrisey, J., & Heidkamp, C.P. (2019), The Blue Economy, identifying geographic concepts and sensitivities, *Geography Compass*, https://doi.org/10.1111/gec3.12445

Graziano, M., Alexander, K., Liesch, M., Lema, E., & Torres, J.A. (2019), Understanding an emerging economic discourse through regional analysis: Blue Economy clusters in the U.S. Great Lakes basin, *Applied Geography*, 105 (2019), 111–123.

Habitat for Humanity (2021), Hurricane Maria Recovery, https://www.habitat.org/impact/our-work/disaster-response/hurricanes/hurricane-maria, last accessed January 25, 2021.

Inter-American Development Bank (IDB) (2021), About, https://www.sustainableislandsplatform.org/about/, last accessed January 25, 2021.

International Alliance of Inhabitants (IAI) (2018), Land grab looms in hurricane-wrecked Barbuda, and what is taking shape is not an airport, https://por.habitants.org/news/habitantes_das_americas/land_grab_looms_in_hurricane-wrecked_barbuda_and_what_is_taking_shape_is_not_just_an_airport, last accessed January 25, 2021.

Latinx Today (2019), Una cuarta parte viviendas en Culebra y Viejo San Juan se alquilan por Airbnb, http://www.latinxtoday.com/6467_ultima-hora/5952487_una-cuarta-parte-viviendas-en-culebra-y-viejo-san-juan-se-alquilan-por-airbnb.html, last accessed January 25, 2021.

NPR (2020), Hurricane Irma May Have Destroyed Barbuda's Generations-Old Land System, https://www.npr.org/2020/02/07/803677660/hurricane-irma-may-have-destroyed-barbudas-generations-old-land-system, last accessed January 2025.

Nuestro Puerto Rico (2021), San Juan, http://nuestropuertorico.com/municipios/san-juan/, last accessed January 25, 2021.

Nweke, O., & Lee, C. (2011), Achieving Environmental Justice: Perspectives on the path forward through collective action to eliminate health disparities, *American Journal of Public Health*, 101, S6–S8.

Research and Statistics Department/Ministry of Tourism, Foreign Affairs and Immigration of Antigua and Barbuda (2017), https://statistics.gov.ag/wp-content/uploads/2018/12/2017-Tourism-Year-in-Review.pdf, last accessed January 25, 2021.

Rawls, J. (1999), *A Theory of Justice* (2nd ed.), Cambridge, MA: Harvard University Press.

Sabbagh, C., & Smith, M. (Eds.) (2018), *Handbook of Social Justice Theory and Research*, New York: Springer.

The Conversation (2019), Barbudans are resisting 'disaster capitalism', two years after hurricane Irma, https://theconversation.com/barbudans-are-resisting-disaster-capitalism-two-years-after-hurricane-irma-119368, last accessed January 25, 2021.

Tourism Bonaire (2021), Blue destination, https://www.tourismbonaire.com/bonaire-blue-destination, last accessed January 25, 2021.

10

CONCEPTUALISING ENTANGLED BLUE ECONOMY AND MARINE SPATIAL PLANNING

Netting Blue Growth and Sustainable Seas in the UK

Gordon M. Winder

10.1 Introduction

Between 2008 and 2014, The European Union (EU) introduced Marine Spatial Planning (MSP) and Blue Economy (BE) frameworks to facilitate the development of marine capabilities for its Blue Growth and Sustainable Seas aspirations. The EU's BE framework supports Blue Growth opportunities and encourages sustainable development with due attention to environmental impacts. Under its ecosystem-based MSP frame, the EU expected rational spatial planning to reduce use conflicts related to intensifying the use of marine resources. These frameworks are being made and entangled, but their effects remain uncertain. BE investment projects are announced as MSP continues to take shape. Case studies of the interaction of BE and MSP making in diverse and politicised marine and coastal spaces are needed. We need to better understand how projects are assessed within these frameworks with respect to economic, social, and environmental outcomes. Geographers and scholars across the marine social sciences are critically engaged with these frameworks but, generally, researchers focus on each framework separately. This paper contends that BE and MSP should be considered together.

The UK has been a leader in the institutionalisation of these frameworks but, after more than a decade of planning, their geographical outcomes remain under-researched. Definitions, rules, and tools are being assembled. While the dynamics of marine clusters have been researched, for example, in Canada (Doloreux et al., 2016) and Ireland (Morrissey, 2015; Morrissey and O'Donoghue, 2012, Morrissey and O'Donoghue, 2013; Morrissey et al., 2011), in the UK, research has tended to focus on renewable energy transition making at the regional scale without much reference to BE (Dawley, 2014; Dawley et al., 2015, 2019; Essletzbichler, 2012; Munro, 2018). As EU priorities shift from realising

DOI: 10.4324/9781003280248-13

Blue Growth to building coastal resilience, managing sustainability transitions, and coping with threats of EU disintegration, it is urgent to research how BE and MSP are being made and with what effects.

Assessments of the achievements and impacts, strengths and weaknesses of BE and MSP planning are overdue. We need to ask whether BE and MSP are netting Blue Growth outcomes for clusters, regions, society, and environments, and whether they are helping to achieve sustainability transition and integration goals. Research should combine analysis of economic growth through marine cluster development associated with BE projects with analysis of environmental outcomes, trade-offs, and bargaining related to these projects and MSP. To what extent are BE and MSP entangled, producing incompatible results; effectively developing forms of economic rationalisation; promoting sustainability transition outcomes; or side-lined by emerging priorities related to climate change, austerity, investment, or nationalism?

This chapter adds to two recent appeals. Garland et al. (2019) called for consideration of BE in regional context using multi-scalar and relational thinking with the aim of achieving just transitions. Bennett (2019a, 2019b) appealed for research on MSP using a political ecology approach and acknowledging contestation and social interests in coastal and ocean environments. In this chapter, I urge assessment of the economic, social, and environmental outcomes of entangled BE and MSP projects. By considering BE and MSP together rather than separately, the extent and consequences of their entanglement can be identified. This will require systematic research of regional and case study experiences of plan making. BE and MSP investments and outcomes should be mapped against goals: are effective marine clusters being built; are social deprivation, inequality, and unemployment being reduced; are environmental objectives being achieved? Can the practices of BE and MSP cope with emerging transition expectations related to making an energy transition, reducing greenhouse gas emissions, building resilient coasts, ecosystems, and communities (Heidkamp and Morrissey, 2019; Turnpenny and O'Riordan, 2007), and effecting integration across boundaries? With this agenda in mind, this paper explores the literature on BE and MSP to establish the state of the art.

10.2 Blue Economy

BE is a form of growth politics closely related to Blue and Green Growth. BE imagining inspires optimism and anticipation. BE reframes natural resources, identifying economic opportunities to be allocated and managed as property under market mechanisms. Integrated governance of uses in marine space is promoted along with investments in infrastructure, marine clusters, and innovation. At a time when governments are confronting multiple and overlapping uses in ocean and marine environments, it is urgent to find ways to avoid collisions in coastal and marine space. BE is also meant to deliver beneficial environmental and developmental futures and to "save" the world's oceans and water.

The EU is a prominent agent promoting BE (Winder and Le Heron, 2017b) and UK policies have been embedded in EU frameworks. BE research supports EU Sustainable Seas policies and EC priorities for Blue Growth, which are buttressed by EU investments.

BE opens the ocean, coast, and interior to new mapping projects and promises to develop economic, environmental, and conservation frontiers. National economies are repartitioned according to whether their economies are ocean or land based. Ocean-based economies are then promoted using cluster initiatives, and policies and projects aimed at infrastructure, research, and development, regrouped economic sectors, or environmental management (see Klein, Lisdat and Tamásy, (2023) this volume). In the related discourse, Europe's BE projects are strongly associated with sustainable development and involve new institutions for governance, marine reserves, and MSP.

BE requires new concepts and practices. Researchers initially focused on how to define and establish the sectoral strengths of Europe's BE and to know its geography (see Morrissey et al., 2011; Smith-Godfrey, 2016; Winder and Le Heron, 2017a). Sea-dependent value chains are separated from national economies. A new combination of sectors (blue growth, coastal and marine tourism, shipbuilding and repair, fisheries, aquaculture, and transport) is identified, and their economic performance calculated for seas (EC, 2014a, 2014b). Researchers focus on how to define BE, on input-output analysis of the marine sector in national and regional economies, and the emergence, definition, and management of maritime clusters (Dalton et al., 2019; Doloreux, 2017; Doloreux et al., 2016; Koliousis et al., 2017; Morrissey, 2015; Morrissey and O'Donoghue, 2012, Morrissey and O'Donoghue, 2013; Morrissey and Cummins, 2016; Morrissey et al., 2011; Surís-Regueiro et al., 2013). Research often operates at new scales, including upscaling to supra-regional scales (Charles, 2017; Fernndez-Macho et al., 2015, 2016). It highlights the importance of tourism, energy, shipping, and ports relative to fishing and lays the groundwork for measuring the future performance of this economy. Researchers stress EU ambitions to build the world's leading bio-economy (Bell et al., 2018). Marine Protected Areas and MSP figure prominently and hopes are held for integration of the EU's Blue Growth and Social Economy initiatives (Banousis et al., 2016). Trade-offs and positive productivity trends are expected (Burgess et al., 2016; Da-Rocha et al., 2019; van den Burg et al., 2019). While potential for social innovation within marine clusters is noted (Pinto et al., 2015; Soma et al., 2018), BE research is only beginning to deal with cluster dynamics, "path dependency" and "resilience."

In contrast, dynamics feature prominently in economic geography research on regional clusters, particularly since "resilience" emerged as a theme in UK geography (Martin, 2012; Fromhold-Eisebith, 2015). However, little of this work focuses explicitly on the UK's marine economy. Nonetheless, a study of Aberdeen's oil cluster found evidence of firms diversifying their markets as North Sea field output contracted, thus highlighting "learning from the North Sea" and the export potential of knowledge creation (Chapman et al., 2004).

Research on the role of Siemens' offshore wind turbine plant in Humber region acknowledges EU funding and offshore investment in cluster development (Dawley, 2014; Dawley et al., 2015, 2019). Politics plays an important role in regional development, especially in offshore wind energy (OWE) investments (Essletzbichler, 2012). Hesse (2018) reveals complexities in port-city relations, including relations with the wider region, cruise ship activity, and institutions, intermediaries, and knowledge brokers, all relevant when assessing BE growth.

Relating BE to social well-being is a difficult task. As Hardill et al. (2001) made clear, the UK's postindustrial landscapes, postcolonial cultural diversity, and economic and social divisions now include "rust pockets" and other micro geographies of disparities. Inequalities in Europe are persistent and require distributed development policies sensitive to place (Iammarino et al., 2019). The idea that UK coasts could be treated as separate economic regions interacting with marine space and social inequality is not prominent in UK economic geography literature. Coe and Jones' (2010) economic geography of the UK documented a persistent North-South divide and addressed energy and climate change but not BE. Further, Braun et al. (2018) identify a need to relate regional clusters to issues of economic growth and social inequality within environmental economic geography. To substantiate the potential of BE to build social sustainability, Morrissey's (2015) benchmark study of relations between the marine sector and employment and deprivation in England needs to be followed up. Nevertheless, Banousis et al. (2016) recently called for integration of the EU's Blue Growth and Social Economy initiatives. Spatial injustices may follow waterfront and inner-city developments (Doucet, 2010; Olson et al., 2013; Smith and Ferrari, 2012), port restructuring, or infrastructure investments (Greco and Bagnardi, 2018; Greco and Di Fabbio, 2014; Karaliotas, 2017, Majekodunmi et al., 2020).

Regional development planning in the UK faces many challenges. These include institutional capacity problems (Gibbs et al., 2001), new forms of regionalism (Jones and MacLeod, 2004), the complexity of planning for sustainable regional development using transitions policy (Turnpenny and O'Riordan, 2007), the dismantling of the UK's regional development agencies (Pike et al., 2016), and devolution (Bailey and Budd, 2016). The scene is complicated by both EU funding for integration projects (Di Cataldo, 2017), changing UK government priorities, and the unfolding consequences of Brexit. Still, there is little sign of marine resources, planning, or connectivity in analyses of uneven regional development in the UK, and perhaps BE may contribute to uneven development.

BE research pays little attention to the environment, including risks and hazards, wastes, pollution, and over-use (but see Heidkamp and Morrissey, 2019). Environments and ecosystem services rarely feature in Blue Growth initiatives (Jones et al., 2019; Mulazzani and Malorgio, 2017). Challenges confront fully integrated ecosystem assessments of BE projects (Link et al., 2019). Hopes that institutional barriers to ecosystem-based management will be overcome (Link et al., 2017) and calls for "balance" between economic and environmental outcomes (Ellis and Flannery, 2016) remain aspirational.

Sustainability frameworks and environmental systems are seldom integrated into BE literature or practice.

Rather, a "pragmatic approach" (Brewer, 2017; Burgess et al., 2016) dominates the BE literature: integrated planning should be pursued by experts who define objectives, quantify trade-offs, engage stakeholders, measure impact, and design institutions, with an emphasis on trust and cooperation (Pinto et al., 2015). Economic rationalisation is expected (Burgess et al., 2016). For example, fisheries are declared to be relatively unimportant because they score low on Total Factor Productivity (Moore et al., 2016; van den Burg et al., 2019). Data are now being collected on European Small-Scale Fisheries fleets for purposes of total factor productivity accounting for future Blue Growth assessments and trade-offs accounting (Da-Rocha et al., 2019). Even without such data, "spatial allocation" decisions are expected to be made in situations where knowledge is limited (e.g., when dealing with sectors with high value but poor data) and must be reconciled with local and indigenous knowledge as well as ecosystem services using scenarios, trade-off analysis, game theoretic decision rules, and strategic mapping products to achieve multi-objective MSP (Lombard et al., 2019). Rationalisation is a work in progress, in part driven by digitalisation, that will impact businesses beyond the fisheries (Dalton et al., 2019; Klinger et al., 2018; Murray et al., 2018). With some exceptions (Heidkamp and Morrissey, 2019), BE literature makes little reference to the costs of Blue Growth, a situation reminiscent of earlier practices related to assessing efficiencies from mechanisation in the fisheries accompanied by inattention to assessments of regional economic, social, and environmental costs (Pinkerton, 2015; Winder, 2018: 12–13).

Critical geographers call for a rights-based approach that pays attention to social impacts of BE (Garland et al., 2019) and prioritises sustainability transition goals over Blue Growth (Kannen, 2014). Sustainability transition approaches are advocated (Kelly et al., 2019; Heidkamp and Morrissey, 2019). They embrace goal setting, engaging stakeholders, and measuring impacts but, in an effort to counter tendencies in planning for "rationalisation," call for use of a rights-based approach combined with attention to social impacts of BE (Cohen et al., 2019). Transitions should be just (Bennett, 2018), involve local participation, and meet government goals. Others see BE as a governmental project producing and rationalising governable spaces for investments (Choi, 2017), as confounded by persistent use of placeless representations (Germond and Germond-Duret, 2016), or as lacking attention to relational thinking (Foley, 2017), value and materialities (Lewis, 2019), interdependencies or temporalities (Winder and Le Heron, 2017a). In their recent review, Garland et al. (2019) note a general failure to integrate geographical concepts such as space and place, proximity and distance, scale and connection, and relational thinking. They call for engagement with the BE term, concept, practices, and implications, using multi-scalar and relational approaches. Further, if we are to assess BE outcomes then economic geography thinking on cluster dynamics, regional development and economic rationalisation is needed. In the context of the varied investment-risk-opportunity profiles of British coasts and marine areas are BE assessment practices improving UK planning, building effective marine clusters and regional economies, and yielding social and environmental outcomes? Finally, BE practices are not only entangled with these other government policies but also with MSP planning. How does BE shape MSP and how is BE itself conditioned by other institutions and policies? Such an approach recognises the nesting of BE in the politics of planning.

10.3 Marine Spatial Planning

MSP aims to reduce conflicts related to intensified use of maritime resources. It emerged over the last 15 years aided by the EU's embrace of the idea in 2007 (Douvere and Ehler, 2009a, 2009b; Kidd and Ellis, 2012) and is practised around the world. The EU's Marine Strategy Framework Directive 2008 required national assessments, including economic, social, and environmental analysis of pressures and impacts from maritime activities, of marine waters by mid-2012 and national marine plans by 2021. The EU is developing MSP as ecosystem-based planning for coastal, sea, and ocean management guided by ecological principles and monitoring (Knol, 2013).

The UK adopted MSP with both energy (Smith et al., 2012) and divergent discourses (Ritchie, 2014). Following pilot projects for the Irish Sea and Pentland Firth and Orkney waters, a burst of activity 2009–2015 institutionalised MSP. First, the Marine Management Organisation (MMO) was established to coordinate MSP in English waters. Marine Strategy Regulations followed in 2010. The UK Marine Policy Statement 2011 set a framework for preparing MSPs. Fifty Marine Conservation Zones (MCZ) were declared in English and Welsh waters before 2016. MMO established 11 plan areas and is developing plans for each in turn (MMO, 2010). Priority went to the East Inshore and Offshore areas where conservation and EU facilitation of OWE services in Humber's marine cluster coincide (Dalton et al., 2019; Dawley, 2014; Essletzbichler, 2012). Planning turned to South Inshore and Offshore areas, where the EU's port alliances, Blue Growth strategy for 2014–2020, and European Regional Development Fund support fostered cooperation among port authorities in Belgium, England, France, and the Netherlands (Winder and Le Heron, 2017b: 6–7). Under this framework, planning is incomplete, numerous government agencies are involved, and Brexit threatens cooperation with European partners. The Marine (Scotland) Act (2010) and National Marine Plan (2015) created a similar framework for Scotland. Marine Scotland drafts plans for 11 marine regions with Advisory Marine Planning Partnerships to ensure local participation and links to territorial authorities. It first completed plans for Clyde and Shetland Islands, but plan making is ongoing. In comparison, MSP has lagged behind in England's Northwest and North East, in Wales and Northern Ireland but the 2021 deadline is achievable even as Brexit and COVID-19 threaten this EU-sanctioned MSP activity.

While MSP required new administration when HE did not, even so, UK MSP is entangled in BE practices. MSP requires study of economic impacts of MPAs, cruise ship and OWE developments, and marine cluster building related to marine science. MMO provided evidence-based guidelines for marine and terrestrial planners working to "maximise socio-economic benefits" from their planning (MMO, 2010). Defining "coastal communities" for statistical and mapping purposes, the report relates "development typologies" to location quotients for marine sector activities and accounts of future investment. It includes a framework for calculating total factor productivity but cautions "productivity drivers may function less well in coastal areas" (MMO, 2010: 1). MMO expects understanding of investors' needs and of how to solve issues that obstruct investment. Similar work was done on the socio-economic effects of Maritime Coastal Zones (MCZ), so the economic rationalisation logic developed around BE is evident in UK MSP.

Generally, assessments question MSP's purpose (Ellis and Flannery, 2016), calling for research into stakeholder engagement (Pomeroy and Douvere, 2008; Ritchie and Ellis, 2010) and who won and lost in planning (Flannery and Ellis, 2016; Johnsen and Hersoug, 2014), in context (Bennett, 2019a; Bennett, 2019b; Jay et al., 2012). MSP is critiqued for expert capture of ocean management (Tafon, 2018; Tafon et al., 2018). It is creating new organisations, territories, and realities (Peters, 2010), but these come with neoliberal socionatures (Bakker, 2010; Castree and Braun, 2001). Early research recommended official attention to stakeholder engagement (Flannery and Cinnéide, 2008; Gopnik et al., 2012). Now MSP is seen to be embedded in diverse projects (Choi, 2017; Flannery et al., 2018; Winder and Le Heron, 2017a) and in marine spaces charged with conflicting rationalities, perceptions, and values (Essletzbichler, 2012; Jay et al., 2012; Jentoft, 2017; Kannen, 2014).

Researchers emphasise MSP's constructed but contingent nature in practice (Boucquey et al., 2016), challenging claims that MSP is apolitical. Exclusion and non-participation are significant problems (Flannery et al. 2018): Fairbanks et al. (2018) contend that MSP is assembling enclosure of the seas. Reactions in Aotearoa-New Zealand were ambivalent when EU-style MSP was presented as best practice: it was seen as opening a rethinking of values (Lewis, 2019) or as failing to meet local expectations around participation (Peart, 2017). In other contexts (Flannery and Ellis, 2016; Mulazzani and Malorgio, 2017; Tafon, 2018), MSP is found to be "post-political planning" (Swyngedouw, 2009): technocratic planning and consensus-based decision-making, legitimised by participation processes, but in effect marginalising opposition and foreclosing debate about the purpose of planning (Flannery et al., 2018). MSP can facilitate investment in some uses at the expense of others but is unlikely to solve underlying conflicts, including inter-agency tensions and politics (Aschenbrenner and Winder, 2019). Increasingly, tensions among experts and scientists are highlighted (Jay et al., 2016). Biological science is foundational in MSP but must be integrated with economics and other sciences. "Scale mismatches" occur (Alexander and Graziano, 2018). In coastal management, the politics of alternative expertise is apparent: for example, Clarke and Rendell (2015) advocate for place-based rather than habitat-based coastal dune management practices.

MSP outcomes, politics, and narratives vary from place to place, so while MSP may be "post-political planning," environmental bargaining (Affolderbach, 2011; Affolderbach et al., 2012) occurs. In marine and coastal space, attention is turning to environmental problems. The European Plastics Strategy now complicates BE and MSP (Penca, 2018). Coastal resilience is called for (Heidkamp and Morrissey, 2019; Turnpenny and O'Riordan, 2007). "Ocean citizens" (Fletcher and Potts, 2007) are more active. Biodiversity conservation targets are not being met. Late last year reports circulated condemning EU MPAs for failing to protect ocean conservation values (McVeigh, 2020). Moreover, a hardening of marine borders is underway in Europe's seas.

In this context, acknowledging contestation in coastal and ocean environments, emphasising social interests, and adopting a political ecology and social science approach (Bennett, 2019a, 2019b) will yield excellent research results. Systematic mapping of MSP activity is needed, including MCZ creation and related marine cluster development. Attention to whether MSP is post-political planning involving rationalisation, exclusion and non-participation, and investigation of tensions among experts is also needed. But, to this agenda, let us also add investigation of the entanglement of MSP with other planning frameworks and institutions. Such an approach will likely demonstrate that MSP is entangled with BE, is politicised and effective, but generates uneven development and compromised land- and seascapes. It will likely reveal inter-agency and expert rivalries and politics within planning processes. Here concepts like "environmental bargaining," "boundary organisations," and "remapping" may help, and a typology of bargaining situations, winners, and losers may be relevant.

10.4 Charting Course

BE and MSP are entangled governance projects and works in progress, assembled differently in each jurisdiction, and associated with politics and bargaining. They need to be studied together and in place. Assessment of the effects of entangled sustainability frameworks in ocean governance and management must remain a feature of research in the marine social sciences. Bargaining and decision-making take place around these frameworks as well as within them, complicating and destabilising expected results. Assessing BE- and MSP-related developments in terms of environmental impacts, making an energy transition and reducing greenhouse gas emissions, building resilient coasts, ecosystems, and communities, and effecting integration across borders are challenging agenda items. The UK's coastal and marine areas are complex and differentiated contexts for assessing BE and MSP effects, especially when the effects must be understood in terms of land and sea interactions. Nevertheless, such a research agenda has great potential to enrich transition-making literature through attention to entangled planning frameworks and practices.

References

Affolderbach, J., (2011) Environmental bargains: Power struggles and decision making over British Columbia's and Tasmania's old-growth forests. *Economic Geography* 87(2): 181–206.

Affolderbach, J., Clapp, R.A., and Hayter, R., (2012) Environmental bargaining and boundary organizations: Remapping British Columbia's Great Bear Rainforest. *Annals of the Association of American Geographers* 102(6): 1391–1408.

Alexander, K., and Graziano, M., (2018) Marine spatial planning, scale mismatches in a complex (regional) seascape. *Regions Magazine* 307(1): 15–16.

Aschenbrenner, M., and Winder, G.M., (2019). Planning for a sustainable marine future? A critical analysis of Marine Spatial Planning in the German Exclusive Economic Zone of the North Sea. *Applied Geography* 110: 102050.

Bailey, D., and Budd, L. (eds.), (2016) *Devolution and the UK Economy*. London: Rowman and Littlefield International.

Bakker, K., (2010) The limits of 'neoliberal natures': Debating green neoliberalism. *Progress in Human Geography* 34(6): 715–735.

Banousis, D., Kyriazi, Z., and Bourtzis, T., (2016) Social economy as a key factor for enhancing Blue Growth in Greece. A conceptual perspective. *AIMS Environmental Science* 3(4): 815–826.

Bell, J., Paula, L., Dodd, T., Németh, S., Nanou, C., Mega, V., and Campos, P., (2018) EU ambition to build the world's leading bioeconomy: Uncertain times demand innovative and sustainable solutions. *New Biotechnology* 40: 25–30.

Bennett, N.J., (2018) Navigating a just and inclusive path towards sustainable oceans. *Marine Policy* 97: 139–146.

Bennett, N.J., (2019a) In political seas: Engaging with political ecology in the ocean and coastal environment. *Coastal Management* 47(1): 67–87.

Bennett, N.J., (2019b) Marine social science for the peopled seas. *Coastal Management* 47(2): 246–253.

Boucquey, N., Fairbanks, L., St. Martin, K., Campbell, L.M., and McCay, B., (2016) The ontological politics of marine spatial planning: Assembling the ocean and shaping the capacities of 'community' and 'environment'. *Geoforum* 75: 1–11.

Braun, B., Oßenbrügge, J., and Schulz, C., (2018) Environmental economic geography and environmental inequality: Challenges and new research prospects. *Zietschrift für Wirtschaftsgeographie* 62(2): 120–134.

Brewer, J., (2017) Actualizing marine policy engagement. *Dialogues in Human Geography* 7(1): 45–49.

Burgess, M.G., Clemence, M., McDermott, G.R., Costello, C., and Gaines, S.D., (2016) Five rules for pragmatic blue growth. *Marine Policy* 87: 331–339.

Castree, N., and Braun, B. (eds.), (2001) *Social Nature: Theory, Practice and Politics*. Malden: Blackwell.

Chapman, K., MacKinnon, D., and Cumbers, A., (2004) Adjustment or renewal in regional clusters? A Study of diversification amongst SMEs in the Aberdeen oil complex. *Transactions of the Institute of British Geographers NS* 29: 382–396.

Charles, K., (2017) Marine science and blue growth: Assessing the marine academic production of 123 cities and territories worldwide. *Marine Policy* 84: 119–129.

Choi, Y.R., (2017) The Blue Economy as governmentality and the making of new spatial rationalities. *Dialogues in Human Geography* 7(1): 37–41.

Clarke, M.L., and Rendell, H.M., (2015) 'This restless enemy of all fertility': Exploring paradigms of coastal dune management in Western Europe over the last 700 years. *Transactions of the Institute of British Geographers* 40: 414–429.

Coe, N.M., and Jones, A. (eds.), (2010) *The Economic Geography of the UK*. London: Sage Publications.

Cohen, P.J., Allison, E.H., Andrew, N.L., Cinner, J., et al., (2019) Securing a just space for small-scale fisheries in the blue economy. *Frontiers in Marine Science* 6: article 171 (8 pages). doi: 10.3389/fmars.2019.00171.

Da-Rocha, J-M., Guillen, J., and Prellezo, R., (2019) (Blue) Growth accounting in small-scale European Union fleets. *Marine Policy* 100: 200–206.

Dalton, G., Bardocz, T., Blanch, M., Campbell, D., et al., (2019) Feasibility of investment in Blue Growth multiple-use of space and multi-use platform projects: Results of a novel assessment approach and case studies. *Renewable and Sustainable Energy Reviews* 107: 338–359.

Dawley, S., (2014) Creating new paths? Offshore wind, policy activism, and peripheral region development. *Economic Geography* 90: 91–112.

Dawley, S., MacKinnon, D., Cumbers, A., and Pike, A., (2015) Policy activism and regional path creation: The promotion of offshore wind in North East England and Scotland. *Cambridge Journal of Regions, Economy and Society* 8(2): 257–272.

Dawley, S., MacKinnon, D., and Pollock, R., (2019) Creating strategic couplings in global production networks: Regional institutions and lead firm investment in the Humber region, UK. *Journal of Economic Geography* 19: 853–872.

Di Cataldo, M., (2017) The impact of EUObjective 1 funds on regional development: Evidence from the U.K. and the prospect of Brexit. *Journal of Regional Science* 57: 814–839.

Doloreux, D., (2017) What is a maritime cluster? *Marine Policy* 83: 215–220.

Doloreux, D., Shearmur, R., and Figueiredo, D., (2016) Quebec's coastal maritime cluster: Its impact on regional economic development, 2001–2011. *Marine Policy* 71: 201–209.

Doucet, B., (2010) Rich Cities with Poor People: Waterfront Regeneration in the Netherlands and Scotland. Doctoral dissertation, University Utrecht.

Douvere, F., and Ehler, C.N., (2009a). New perspectives on sea management: Initial findings from European experience with marine spatial planning. *Journal of Environmental Management* 90: 77–88.

Douvere, F., and Ehler, C.N., (2009b) Marine Spatial Planning: A Step-by-Step Approach toward Ecosystem-Based Management. Intergovernmental Oceanographic Commission and Man and the Biosphere Programme. IOC Manual and Guides, 53, ICAM Dossier, 6. Retrieved from http://unesdoc.unesco.org/images/0018/001865/186559e.pdf (accessed 20 June 2018).

EC, (2014a) *Blue Growth, Brussels, European Commission. Communication from the Commission to the European Parliament, The Council, The European Economic and Social Committee and the Committee of the Regions EC Blue Growth Year.* Available at: http://eur-lex.europa.eu/legal-content/EN/ALL/?uri=CELEX:52012DC049 (accessed 31 August 2015).

EC, (2014b) *Innovation in the Blue Economy: Realising the Potential of Our Seas and Oceans for Jobs and Growth*. Brussels: COM.

Ellis, G., and Flannery, W., (2016) Marine spatial planning: Cui bono? *Planning Theory and Practice* 17(1): 122–128.

Essletzbichler, J., (2012) Renewable energy technology and path creation: A multi-scalar approach to energy transition in the UK. *European Planning Studies* 20: 791–816.

Fromhold-Eisebith, M., (2015) Sectoral resilience: Conceptualizing industry-specific spatial patterns of interactive crisis adjustment. *European Planning Studies*. doi: 10.1080/09654313.2015.1047329.

Fairbanks, L., Campbell, L.M., Boucquey, N., and St. Martin, K., (2018) Assembling enclosure: Reading marine spatial planning for alternatives. *Annals of the Association of American Geographers* 108(1): 144–161.

Fernández-Macho, J., et al., (2015) Measuring the maritime economy: Spain in the European Atlantic Arc. *Marine Policy* 60: 49–61.

Fernández-Macho, J., González, P., and Virto, J., (2016) An index to assess maritime importance in the European Atlantic economy. *Marine Policy* 64: 72–81.

Flannery, W., and Cinnéide, M.O., (2008) Marine spatial planning from the perspective of a small seaside community in Ireland. *Marine Policy* 32: 980–987.

Flannery, W., and Ellis, G., (2016) Exploring the winners and losers of marine environmental governance. *Planning Theory and Practice* 17(1): 121–122.

Flannery, W., Healy, N., and Luna, M., (2018) Exclusion and non-participation in marine spatial planning. *Marine Policy* 88: 32–40.

Fletcher, S., and Potts, J., (2007). Ocean citizenship: An emergent geographical concept. *Coastal Management* 35: 511–524.

Foley, R., (2017) One blue. *Dialogues in Human Geography* 7(1): 32–36.

Garland, M., Axon, S., Graziano, M., Morrissey, J., and Heidkamp, C.P., (2019) The blue economy: Identifying geographic concepts and sensitivities. *Geography Compass* 2019: e12445. http://doi.org/10.1111/gec3,12445

Germond, B., and Germond-Duret, C., (2016) Ocean governance and maritime security in a placeful environment: The case of the European Union. *Marine Policy* 66: 124–131.

Gibbs, D.C., Jonas, A.E.G., Reimer, S., and Spooner, D.J., (2001) Governance, institutional capacity and partnerships in local economic development: Theoretical issues and empirical evidence from the Humber Sub-region. *Transactions of the Institute of British Geographers NS* 26: 103–119.

Gopnik, M., Fiesler, C., Cantral, L., McClellan, K., Pendleton, L., and Crowder, L., (2012) Coming to the table: Early stakeholder engagement in marine spatial planning. *Marine Policy* 36(5): 1139–1149.

Greco, L., and Di Fabbio, M., (2014) Path-dependence and change in an old industrial area: The case of Taranto, Italy. *Cambridge Journal of Regions, Economy and Society* 7(3): 413–431.

Greco, L., and Bagnardi, F., (2018) In the name of science and technology: The post-political environmental debate and the Taranto Steel Plant (Italy). *Environmental Values* 27(5): 489–512.

Hardill, I., Graham, D.T., and Kofman, E., (2001) *Human Geography of the UK: An Introduction*. London and New York: Routledge.

Heidkamp, C.P., and Morrissey, J.E., (eds.), (2019) *Towards Coastal Resilience and Sustainability*. London and New York: Routledge.

Hesse, M., (2018) Approaching the relational nature of the port-city interface in Europe: Ties and tensions between seaports and the urban. *Tijdschrift voor Economische en Sociale Geografie* 109(2): 210–223.

Iammarino, S., Rodriguez-Pose, A., and Storper, M., (2019) Regional inequality in Europe: Evidence, theory and policy implications. *Journal of Economic Geography* 19 (2019): 273–298.

Jay, S., Klenke, T., Ahlhorn, F., and Ritchie, H., (2012) Early European experience in marine spatial planning: Planning the German exclusive economic zone. *European Planning Studies* 20(12): 2013–2031.

Jay, S., Klenke, T., and Janßen, H., (2016) Consensus and variance in the ecosystem approach to marine spatial planning: German perspectives and mutli-actor implications. *Land Use Policy* 54: 129–138.

Jentoft, S., (2017) Small-scale fisheries within marine spatial planning: Knowledge integration and power. *Journal of Environmental Policy and Planning* 19(3): 266–278.

Johnsen, J.P., and Hersoug, B., (2014) Local empowerment through the creation of coastal space? *Ecology and Society* 19(2). https://doi.org/10.5751/ES-06465-190260

Jones, M., and MacLeod, G., (2004) Regional spaces, spaces of regionalism: Territory, insurgent politics and the English question. *Transactions of the Institute of British Geographers NS* 29: 433–452.

Jones, D.O.B., Durden, J.M., Murphy, K., Gjerde, K.M., et al., (2019) Existing environmental management approaches relevant to deep-sea mining. *Marine Policy* 103: 172–181.

Kannen, A., (2014) Challenges for marine spatial planning in the context of multiple sea uses, policy arenas and actors based on experiences from the German North Sea. *Regional Environmental Change* 14(6): 2139–2150.

Karaliotas, L., (2017) Performing neoliberalization through urban infrastructure: Twenty years of privatization policies around Thessaloniki's port. *Environment and Planning A: Economy and Space* 49(7): 1556–1574.

Kelly, C., Ellis, G., and Flannery, W., (2019) Exploring transition pathways as an alternative approach for the integrated management of Irish estuaries and coasts. In C.P. Heidkamp, and J.E. Morrissey (eds.), *Towards Coastal Resilience and Sustainability*. London and New York: Routledge: 81–102.

Kidd, S., and Ellis, G., (2012) From the land to the sea and back again? Using terrestrial planning to understand the process of marine spatial planning. *Journal of Environmental Policy and Planning* 14(1): 49–66.

Klein, O. Lisdat, C., and Tamásy, C. (2023). Blue Economy agenda for the Baltic Sea region. In C.P Heidkamp, J. Morrissey, C. Germond-Duret (Eds.), *Blue Economy: People and regions in transitions*. Oxford, UK: Routledge.

Klinger, D.H., Eikeset, A.M., Davidsðóttir, B., Winter, A.-M., and Watson, J.R., (2018) The mechanics of blue growth: Management of oceanic natural resource use with multiple, interacting sectors. *Marine Policy* 87: 356–362.

Knol, M., (2013) Making ecosystem-based management operational: Integrated monitoring in Norway. *Maritime Studies* 12(1): 5.

Koliousis, I.G., Papadimitriou, S., Riza, E., Stavroulakis, P.J., and Tsioumas, V., (2017) Strategy, policy and the formulation of maritime cluster typologies. *Marine Policy* 86: 31–38.

Lewis, N., (2019) Cultivating diverse values by rethinking blue economy in New Zealand. In C.P. Heidkamp and J. Morrissey (eds.), *Towards Coastal Resilience and Sustainability*. London and New York: Routledge.

Link, J.S., Dickey-Collins, M., Rudd, M., et al., (2019) Food for thought: Clarifying mandates for marine ecosystem-based management. *ICES Journal of Marine Science* 76(1): 41–44.

Link, J.S., Thébaud, O., Smith, D.C., et al., (2017) Introduction: Keeping humans in the ecosystem. *ICES Journal of Marine Science* 74(7): 1947–1956.

Lombard, A., Ban, N.C., Smith, J.L., et al., (2019) Practical approaches and advances in spatial tools to achieve multi-objective Marine Spatial Planning. *Frontiers in Marine Science* 6: article 166 (9 pages).

MacKinnon, D., Dawley, S., Steen, M., Menzel, M. P., Karlsen, A., Sommer, P., Hansen, G. H., and Normann, H.E., (2018) Path creation, global production networks and regional development: A comparative international analysis of the offshore wind sector. *Progress in Planning 30*: 1–32.

McVeigh, K., (2020) Auditors slam EU for 'marine protected areas' that fail to protect ocean. *Guardian*, 3 December.

Majekodunmi, M., Emmanuel, R., and Jafry, T., (2020). A spatial exploration of deprivation and green infrastructure ecosystem services within Glasgow city. *Urban Forestry & Urban Greening*, 52(2020), doi: 126698.

Martin, R., (2012) Regional economic resilience, hysteresis and recessionary shocks. *Journal of Economic Geography* 12: 1–32.

MMO, (2010) *Maximising the Socio-economic Benefits of Marine Planning for English Coastal Communities* (July 2011). https://assets.publishing.service.gov.uk/government/uploads/system/uploads/attachment_data/file/312720/se_national.pdf

Moore, F., Lamond, J., and Appleby, T., (2016) Assessing the significance of the economic impact of Marine Conservation Zones in the Irish Sea upon the fisheries sector and regional economy in Northern Ireland. *Marine Policy* 74: 136–142.

Morrissey, K., (2015) An inter and intra-regional exploration of the marine sector employment and deprivation in England. *The Geographical Journal* 181(3): 295–303.

Morrissey, K., and Cummins, V., (2016) Measuring relatedness in a multi-sectoral cluster: An input-output approach. *European Planning Studies* 24(4): 629–644.

Morrissey, K., and O'Donoghue, C., (2012) The Irish marine economy and regional development. *Marine Policy* 36(2): 358–364.

Morrissey, K., and O'Donoghue, C., (2013) The role of the marine sector in the Irish national economy: An input-output analysis. *Marine Policy* 37: 230–238.

Morrissey, K., O'Donoghue, C., and Hynes, S., (2011) Quantifying the value of multi-sectoral marine commercial activity in Ireland. *Marine Policy* 35(5): 721–727.

Mulazzani, L., and Malorgio, G., (2017) Blue growth and ecosystem services. *Marine Policy* 85: 17–24.

Munro, F.R., (2018). Renewable energy and transition-periphery dynamics in Scotland. *Environmental Innovation and Societal Transitions* 31: 273–281.

Murray, F., Needham, K., Gormley, K., Rouse, S., et al., (2018) Data challenges and opportunities for environmental management of North Sea oil and gas decommissioning in an era of blue growth. *Marine Policy* 97: 130–138.

Olson, E., Hopkins, P., Pain, R., et al., (2013) Variations of the entrepreneurial city: Glasgow harbor mega-projects. *International Journal of Urban and Regional Research* 37(6): 2035–2051.

Peart, R., (2017) A 'sea change' in marine planning: The development of New Zealand's first marine spatial plan. *Policy Quarterly* 13(2): 3–9.

Penca, J., (2018) European Plastics Strategy: What promise for global marine litter? *Marine Policy* 97: 197–201.

Peters, K., (2010) Future promises for contemporary social and cultural geographies of the sea. *Geography Compass* 4(9): 1260–1272.

Pike, A., Coombes, M., O'Brien, P., and Tomaney, J., (2016) Austerity states, institutional dismantling and the governance of sub-national economic development: The demise of the regional development agencies in England. *Territory, Politics, Governance* 6(1): 118–144.

Pinkerton, E., (2015). Groundtruthing individual transferable quotas. In E.P. Durrenberger and G. Palsson (eds.), *Gambling Debt: Iceland's Rise and Fall in the Global Economy*. Boulder, CO: University Press of Colorado: 109–120.

Pinto, H., Cruz, A.R., and Combe, C., (2015) Cooperation and the emergence of maritime clusters in the Atlantic: Analysis and implications of innovation and human capital for blue growth. *Marine Policy* 57: 167–177.

Pomeroy, R., and Douvere, F., (2008) The engagement of stakeholders in the marine spatial planning process. *Marine Policy* 32: 816–822.

Ritchie, H., (2014) Understanding emerging discourses of marine spatial planning in the UK. *Land Use Policy* 38: 666–675.

Ritchie, H., and Ellis, G., (2010) 'A system that works for the sea'? Exploring stakeholder engagement in marine spatial planning. *Journal of Environmental Planning and Management* 53(6): 701–723.

Smith, H., Ferrari, G. (eds.), (2012) *Waterfront Regeneration. Experiences in City-building.* Abingdon: Routledge.

Smith, H.D., Ballinger, R.C., and Stojanovic, T.A., (2012) The spatial development basis of marine spatial planning in the United Kingdom. *Journal of Environmental Policy and Planning* 14(1): 29–47.

Smith-Godfrey, S., (2016) Defining the blue economy. *Maritime Affairs: Journal of the Maritime Foundation of India* 12(1): 58–64.

Soma, K., van der Burg, S.W., Hoefnagel, E.W., Stuiver, M., and van der Heide, C.M., (2018) Social innovation: A future pathway for blue growth? *Marine Policy* 87: 363–370.

Surís-Regueiro, J.C., Garza-Gil, M.D., and Varela-Lafuente, M., (2013) Marine economy: A proposal for its definition in the European Union. *Marine Policy* 42: 111–124.

Swyngedouw, E., (2009) The antimonies of the postpolitical city: In search of a democratic politics of environmental production. *International Journal of Urban and Regional Research* 33(3): 601–620.

Tafon, R.V., (2018) Taking power to sea: Towards a post-structuralist discourse theoretical critique of marine spatial planning. *Environment and Planning C: Politics and Space* 36(2): 258–273.

Tafon, R.V., Howarth, D., and Griggs, S., (2018) The politics of Estonia's offshore wind energy programme: Discourse, power and marine spatial planning. *Environment and Planning C: Government and Policy* 37(1): 157–176.

Turnpenny, J., and O'Riordan, T., (2007) Editorial: Putting sustainability science to work: Assisting the East of England Region to respond to the challenges of climate change. *Transactions of the Institute of British Geographers NS* 32: 102–105.

van den Burg, S.W.K., Aguilar-Manjarrez, J., Jenness, J., and Torrie, M., (2019) Assessment of the geographical potential for co-use of marine space, based on operational boundaries for Blue Growth sectors. *Marine Policy* 100: 43–57.

Winder, G.M. (ed.), (2018). *Fisheries, Quota Management and Quota Transfer: Rationalization through Bio-economics.* Berlin: Springer.

Winder, G.M., and Le Heron, R., (2017a). Assembling a Blue Economy moment? Geographic engagement with globalising biological-economic relations in multi-use marine environments. *Dialogues in Human Geography* 7(1): 3–26.

Winder, G.M., and Le Heron, R., (2017b). Further assembly work: A mountains to seas Blue Economy imaginary. *Dialogues in Human Geography* 7(1): 50–55 (response to commentaries).

11

BLUE ECONOMY AGENDA FOR THE BALTIC SEA REGION

Oliver Klein, Clemens Lisdat, and Christine Tamásy

11.1 Introduction

Despite a growing interest in the Blue Economy worldwide, the concept embraces diverse strategies and practices to manage marine and coastal regions. Research activities on the Blue Economy have been diverse too, as illustrated by Garland et al. (2019) who state furthermore that researchers, policymakers, and practitioners often ignore geographical concepts, when promoting economic development in such environments. In this chapter, we use the Baltic Sea Region (BSR) in northern Europe as an example for a Blue Economy agenda "in the making." Surrounded by nine countries with diverse political, social, and economic realities, the BSR is governed by different actors and interests that are connected through EU-wide, national, regional, or local policy initiatives. This constellation may allow for critical insights on strategic framings, governance mechanisms, and discursive constructions of sustainability within a regional Blue Economy setting.

Socio-scientific research on the BSR in combination with Blue Economy framings has been relatively scarce so far. However, from our literature review, we could find two significant directions. First, a number of studies focus on maritime spatial planning (MSP) with special attention of transnational integration, coordination, and management practices (Hassler et al., 2019; Morf et al., 2019; Westholm, 2018). Second, there is major interest in ecosystem-based management as a specific form of environmental governance (Söderström & Kern, 2017; Hegland et al., 2015). As these studies mostly rely on practical approaches, we recognise a lack of theory-driven research on Blue Economy strategies at the regional scale. Moreover, there is little research on holistic BSR development based on an integrated analysis of different policy frameworks. At least, some authors are engaged with the composition and impact of single strategies

DOI: 10.4324/9781003280248-14

or action plans (Ollikainen et al., 2019; Gänzle & Kern, 2016). In fact, however, there are still some research gaps left, especially regarding the meaning(s) of sustainability and its realisation within the emerging Blue Economy agenda for the BSR.

Addressing these research gaps, our chapter sheds light on recent developments in governing the Blue Economy agenda for the BSR from a human geography perspective. With a critical gaze, we look at the most important policies and events, while questioning the composition of economic, environmental, and societal relations in strategic framing of the Blue Economy. Following Winder and Le Heron (2017), we use "assemblage thinking" to conceptualise our research based on content analysis of recent academic literature, policy reports, and strategic policy documents. Following the introduction, Section 11.2 outlines the idea of Blue Economy assemblages as conceptual backbone of this chapter. Section 11.3 provides a mapping of policies, networks, and events that are critical components of the emerging Blue Economy in the BSR, before analysing selected findings through the lens of assemblage thinking. In Section 11.4, we provide a critical outlook including implications for future research.

11.2 Blue Economy Assemblages

Assemblage means an unpredictable, fluid, short-time existing, yet vital and emotional set of heterogeneous components (persons or things; Müller & Schurr, 2016). An assemblage does not determine or exquisite forces. It is a space of always changing opportunities towards the perceived and promoted norms of the world (Savage, 2019). The fluctuating process of becoming, existing, and passing of a specific assemblage is central. Focus lies upon possibilities and circumstances of grouping assemblages, while separating into various related components (Müller & Schurr, 2016; Anderson et al., 2012). However, no welding of components to fixed and normed entities occurs. A relation between different components may happen but is not essential for their existence according to assemblage thinking. McFarlane and Anderson (2011) stress the importance of a broad view, when using assemblage thinking in human geography. This way of thinking is helpful, first, to analyse how different historical relations came into being. Second, as an ethical orientation, it provides guidance within the realm of instable interactions, the synchronous possibility of something new and of space-time distinctions. Third, as a concept, assemblage thinking illustrates the relationship between stability and transformation in the process of social formation (Anderson et al., 2012).

In line with the above arguments, we use assemblage theory as a way of thinking about the Blue Economy (Baker & McGuirk, 2017), guiding research methodologically as an "act of problematisation" (Marks, 2005, p. 284). Thinking through assemblage exceeds pure description by elaborating the world through analytical interpretations (Duffy & Stojanovic, 2018). According to assemblage theory, a blueprint for research does not exist, stating exactly which materials and way of data collection are best. But assemblage thinking is suspicious against language, which encounters ontological differentiations (e.g. women/men), while neutrality is to deny. Language is influenced by and is influencing materiality and vice versa. At the same time, both do follow an intrinsic logic. Only within the context of an assemblage, language has performative power. Material or symbolic entities give the power over symbols or material to the "word" (Mattissek & Wiertz, 2014). Ontology, therefore, is not the question of genesis, but alienation forms common representation and the alienation of understanding is pre-existing. In doing so, issues emerge in a positive way (Kristensen et al., 2014), focusing on how material and symbol work together to change the balance of power.

Power is in constant flux with multiple centres simply "everywhere" (Savage, 2019). Fairbanks et al. (2018) convincingly relate assemblage to enclosure, while the latter connects to governmentality and the power to control. Due to the impropriety of objectified denominations within assemblage thinking, policy is no certainty. Policy to foster the Blue Economy is instead a factual affair, only possible to decode within the contingent, changeable composing parts by which the policy is generated (Savage, 2019). Accordingly, policy focuses upon the relative possibility and achievements of persons (individuals) or organisations in performing power. Policy, therefore, is a process with the aim to direct somebody (individuals or groups) in a wanted direction, converting ideas and practice (Fairbanks et al., 2018). States exert power but do not hold it exclusively. States somehow gamble with power in an effort to guide behaviour in and beyond territorial claims (Savage, 2019).

Utilising assemblage thinking for Blue Economy analyses means to explore their inherent meanings, practices, and challenges as well as human and non-human entanglements. Therefore, assemblage thinking may help revealing "the multiple ontologies of investment institutional initiatives and the particular economic-environment relations invoked, established and stabilized with Blue Economy imagining" (Winder & Le Heron, 2017, p. 4). It is important to note, however, that Blue Economy assemblages also involve a regional context dimension that can only be grasped in consideration of associated multi-scalar socio-ecological complexities (Garland et al., 2019). This form of relational thinking may contribute to better understanding the emergence of Blue Economy policies, as it simultaneously points to the diversity of "desires" and interests coming together to achieve particular, issue-related goals in coastal and marine development. Thus, diverse actualisations of governance or, in other words, *governance assemblages* (Briassoulis, 2019) do necessarily co-exist in the Blue Economy and its different regional manifestations. Following from these reflections, the Blue Economy may be best interpreted as an "opportunity for assembling separate marine projects in multi-use and multi-user spaces" (Winder & Le Heron, 2017, p. 20). These projects are very heterogenous as they include, or refer to, marine policies, climate change, exclusive economic zones, marine protected areas, environmental monitoring, stock assessments, aquaculture licenses, port developments, energy infrastructure, transport routes, tourist attractions, microplastics, and so forth.

11.3 Baltic Sea Region: Towards a Sustainable Blue Economy?

In the European Union (EU), the development of marine and coastal spaces has been accelerated by targeted policies beginning in the second half of the 2000s. The political debates were strongly driven by economic growth imperatives, leading to the formulation of the EU Blue Growth vision in 2012. Established as the maritime pillar of the Europe 2020 strategy, the aim of Blue Growth was to achieve a "smart, sustainable and inclusive economic and employment growth from the oceans, seas and coasts" (ECORYS et al., 2012, p. 26). The implementation of Blue Growth was, and still is, closely linked with MSP and several related maritime policies, while also acknowledging the regional differences of marine and coastal spaces across the EU.

The BSR is located in northern Europe and consists of nine countries with different cultures and histories (see Figure 11.1). In May 2014, the European Commission (EC) adopted the Sustainable Blue Growth Agenda for the BSR, which however has become subject to criticism. For example, Ertör & Hadjimichael (2020, p. 3) suggest that "discussions over the future of the ocean are in need of

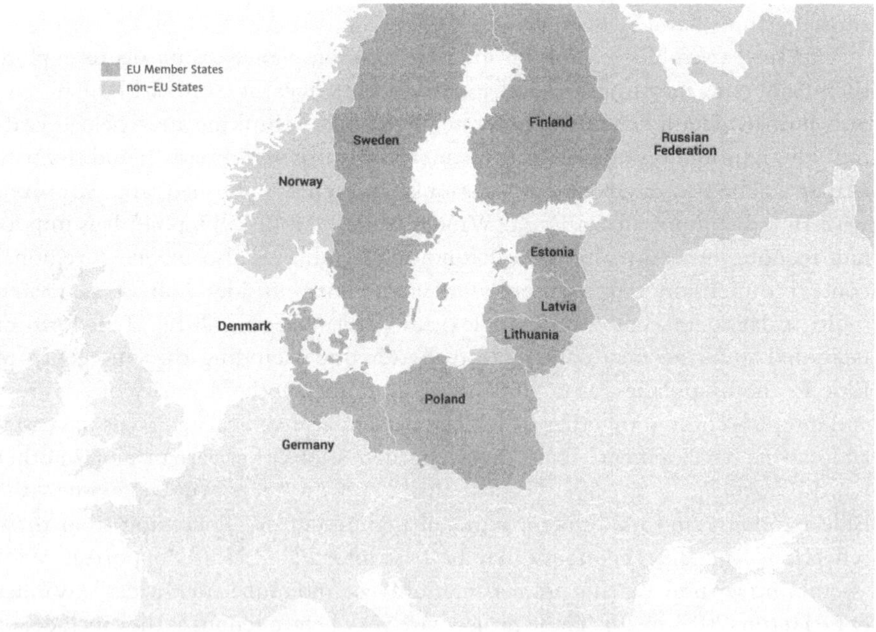

FIGURE 11.1 Baltic Sea Region

a more radical and critical stance challenging the mainstream discussions fixed on the present growth-oriented imaginary." In face of such criticism, the EC has recently put forward the notion of Blue Economy in order to emphasise the goal of *sustainable* economic development of marine and coastal spaces (EC, 2020a). We take this strategic (or rhetoric?) move as a point of departure to critically analyse recent manifestations of the Blue Economy in the BSR from an assemblage perspective. We do so by tracing and studying relevant policies, networks, and events, before linking the findings with conceptual thoughts.

The selection of materials was driven by pertinent references made in academic literature in combination with extensive search via Google. Accordingly, we selected policies, networks, and events that are intended to further develop, transform, or integrate the BSR as a whole. This particularly applies to policy frameworks initiated by the EU, providing overarching strategies, visions, goals, and fields of action in order to frame a (supposed) future-oriented Blue Economy agenda. Furthermore, we included policies that explicitly refer to the notion of "Blue Economy" in a BSR context. The selection of networks and events—aside from their strategic importance—also took into account their continuity over many years as well as their transnational and integrative nature.

11.3.1 Policies, Networks, Events

The strategic development of the BSR got significant impetus by introduction of the EU Strategy for the Baltic Sea Region (EUSBSR), which pursues three objectives that are still current: (1) saving the sea, (2) connecting the region, and (3) increasing prosperity (EC, 2020b). In order to achieve regional development in a structured manner, the EUSBSR Action Plan is subdivided into 14 Policy Areas covering 44 Actions (see Figure 11.2). The practical implementation of this framework rests, among others, on so-called flagships taking the form of projects (single or grouped in clusters) or processes (e.g. networks, cooperation platforms). As such, flagships demonstrate the progress of the EUSBSR (EC, 2020b). In order to initiate and strengthen collaboration among stakeholders, the EUSBSR Annual Fora have been established as a meeting place for actors, organisations, and projects. Given its comprehensive nature, detailed structure, and broad legitimation, the EUSBSR is the overarching foundation of further policies aiming at BSR development.

The aforementioned Sustainable Blue Growth Agenda is closely linked with the EUSBSR. In general, the agenda aims to identify "specific characteristics and strengths through which the full potential for growth and innovation of the region's maritime economies could be harnessed" (EC, 2014, p. 1). In order to realise and further develop Blue Growth, a systematic stakeholder dialogue has been launched by the EC in 2016, focusing on four high-potential thematic areas: (1) shipping, (2) blue bioeconomy (including aquaculture), (3) coastal and maritime tourism, and (4) environmental and monitoring technology. The definition of these areas marks a revision of the original draft of the framework that

FIGURE 11.2 Objectives and policy areas of EUSBSR

included five key sectors. The findings of the stakeholder dialogue were summarised in the form of "strategic transformation maps" for each of the four thematic areas, thus providing a guide to translate Blue Growth into specific action and project work (s.Pro, 2017).

Fostering transnational projects and cooperation, the Interreg Baltic Sea Region Programme plays an important role for integrated BSR development. Funded by the EU and approved by the EC, the programme is targeted at four thematic priorities: (1) capacity for innovation, (2) management of natural resources, (3) sustainable transport, and (4) EU strategy support (EU, 2020a). With regard to the imminent funding period 2021 to 2027, the EC (2019) has recently issued an Orientation Paper, in which ideas, options, and guiding principles are presented. Accordingly, a special focus should be laid on the "Save the sea" objective of the EUSBSR, while simultaneously limiting support for innovation activities. This recommendation is reasoned by the worrying environmental status of the Baltic Sea and the increasing impact of climate change (EC, 2019).

Another policy refers to the SUBMARINER Network, which was initiated in 2013 as a flagship project within EUSBSR. Following the goal of actively promoting innovative and sustainable uses of marine resources, the network regards itself as representing the Blue Economy in the BSR (SUBMARINER Network for Blue Growth EEIG, 2018). The policy guideline of the network is the SUBMARINER Roadmap based on a concrete vision "towards a blue-green economy." This vision incorporates natural capital preservation, sustainable energy and biomass production, human well-being as well as bio-based innovations and smart combinations. In order to specify these goals, a number of strategic action fields (e.g. energy, ecosystem services, blue biotechnology) have been identified (SUBMARINER Network for Blue Growth EEIG, 2015).

A more explicit focus on the Blue Economy is taken by the World Wildlife Fund (WWF) under the umbrella of its Baltic Ecoregion Programme. Following the argument that "a healthy sea is vital to a healthy economy," the WWF presents a set of Blue Economy principles to guide development and assesses the BSR's status and trajectory relative to these principles by means of a SWOT analysis (WWF, 2015). In the conclusion, the WWF draws a visionary picture: "Imagine a Blue Economy that, as it grows and develops, actually helps bring the Baltic Sea back to health, by fitting within the boundaries of the sea's ecosystems, accelerating the adoption of clean technologies and renewable energy, and creating circular material flows. (...)" (WWF, 2015, p. 30).

In a similar vein are the activities of the so-called Helsinki Commission (HELCOM). A recent milestone in the long history of HELCOM was the release of a status quo report on the Baltic Sea Action Plan (BSAP) (HELCOM, 2018). The overarching goal of the BSAP, which was originally adopted in 2007, is to reach good environmental status of the Baltic Sea by 2021. This ambition is specified by four sub-goals: (1) Baltic Sea unaffected by eutrophication, (2) Baltic Sea undisturbed by hazardous substances, (3) environmentally friendly maritime activities, and (4) favourable status of Baltic Sea biodiversity (HELCOM, 2018). For each sub-goal, a number of more targeted aims and actions—most of them being linked to the field of biodiversity—are formulated. The implementation of the BSAP is particularly based on innovative ecological modelling and assessment techniques.

There are a number of further policies and networks in the BSR that have developed over time (see Table 11.1). From a political viewpoint, the Council of the Baltic Sea States (CBSS) is the most important institution. Established in 1992, the CBSS still provides a political forum for regional cooperation at nation-state level. Other networks to be mentioned are the Baltic Development Forum, serving as a think-tank to advance growth and competitiveness in the BSR from 1998 to 2018, and the Baltic SCOPE project, representing a targeted initiative relating to MSP.

The policies and networks presented indicate that joint events play a major role in shaping a Blue Economy in the BSR (see Table 11.2). At this point, we only provide an overview of selected events (from 2015 to 2020) that we think are of greater importance for recent implementation of macro-regional policies. In this respect, the EUSBSR Annual Forum clearly acts as a key event due to its continuous purpose of facilitating intense dialogue among stakeholders (EU,

TABLE 11.1 Blue Economy policies in the BSR

Actor(s)	*Strategy*	*Main documents*	*Thematic focus/goals*
European Union, European Commission	EU Strategy for the Baltic Sea Region (EUSBSR) (2009)	– Commission Communication concerning the EUSBSR (2009) – Working Document: Roles and responsibilities of the implementing stakeholders of the EUSBSR and a flagship project concept (2013) – Working Document: EUSBSR Action Plan (2017) – Annex to the EUSBSR Action Plan (2018) – EUSBSR Action Plan Agreed by the NC Group (2020)	Sustainable macro-regional development (= overarching strategic framework for the BSR) – Saving the sea – Connecting the region – Increasing prosperity – Definition of 14 policy areas covering 44 actions
European Union, European Commission	EU Blue Growth Strategy (2014)	– Working Document: A Sustainable Blue Growth Agenda for the Baltic Sea Region (2014) – Towards an Implementation Strategy for the Sustainable Blue Growth Agenda for the Baltic Sea Region (2017)	Sustainable economic development – Boosting innovation and sustainability – Developing skills and qualifications, cluster development – Using existing cooperation structures and multisectoral dialogue – Targeting maritime projects for access to finance
European Union, European Commission	Interreg Baltic Programme (2014, 2021)	– Interreg BSR Cooperation Programme (2014) – Annual Implementation Reports (2014/15, 2016, 2017, 2018, 2019) – Orientation Paper: Interreg Baltic Sea Region 2021–2027 (2019)	Integrated territorial development and cooperation – Capacity for innovation – Management of natural resources – Sustainable transport – EU Strategy support

(Continued)

TABLE 11.1 Blue Economy policies in the BSR *(Continued)*

Actor(s)	*Strategy*	*Main documents*	*Thematic focus/goals*
Helsinki Commission (HELCOM)	Baltic Sea Action Plan (BSAP) (2007)	– HELCOM Baltic Sea Action Plan (2007) – HELCOM Holistic Assessments of the State of the Baltic Sea (2010, 2017) – Implementation of the Baltic Sea Action Plan (2018)	Sustainable environmental development ("reaching good environmental status") – Baltic Sea unaffected by eutrophication – Baltic Sea undisturbed by hazardous substances – Environmentally friendly maritime activities – Favourable status of Baltic Sea biodiversity
Different stakeholders representing research institutions, public administrations, business parks, and private companies	SUBMARINER Network (2013)	– SUBMARINER Compendium: An Assessment of Innovative and Sustainable Uses of Baltic Marine Resources (2012) – SUBMARINER Roadmap: Towards a Blue-green Economy in the Baltic Sea Region (2015) – SUBMARINER Roadmap Status Report: Towards a Blue-green Economy in the Baltic Sea Region (2016)	Developing a blue-green economy through innovative and sustainable uses of marine resources – Preservation of natural capital – Sustainable energy and biomass production – Improving human well-being – Fostering bio-based innovations and smart combinations
World Wildlife Fund	WWF Baltic Ecoregion Programme (2013)	– Baltic Sea Action Plan—is it on track? (2013) – WWF Report: All Hands on Deck—Setting Course Towards a Sustainable Blue Economy (2015)	Environmental protection and development → Implementation of an ecosystem-based integrated oceans management approach
Responsible authorities for Maritime Spatial Planning from Estonia, Latvia and Sweden	Baltic SCOPE (2015)	– Case Study Report: Towards Coherent Cross-Border Maritime Spatial Planning in the Central Baltic Sea (2017)	Maritime Spatial Planning (MSP) – Increase collaboration between national authorities and other key MSP stakeholders – Finding common approaches to solve transboundary MSP issues – Enhance alignment of national maritime spatial plans

TABLE 11.2 Selected stakeholder events in the BSR (2015–2020)

Date	*Place*	*Event*
18–20 March 2015	St. Petersburg, Russia	16th International Environmental Forum "Baltic Sea Day"
26–27 March 2015	Kiel, Germany	"Blue Growth in the Baltic": Kick-off Conference of Maritime Stakeholder Platform in the Baltic Sea Region
15–16 June 2015	Jūrmala, Latvia	6th EUSBSR Annual Forum
22–23 September 2015	Gdánsk, Poland	8th Baltic Sea Tourism Forum
23–25 September 2015	Visby, Sweden	23rd Baltic Sea States Sub-regional Co-operation (BSSSC) Conference
6 October 2015	Copenhagen, Denmark	Conference "Drivers for Blue Growth in the Baltic Sea Region"
23 November 2015	Copenhagen, Denmark	17th Baltic Development Forum Summit
22–23 March 2016	St. Petersburg, Russia	17th International Environmental Forum "Baltic Sea Day"
14–16 September 2016	Hämeenlinna, Finland	24th Baltic Sea States Sub-regional Co-operation (BSSSC) Conference
19–20 October 2016	Pärnu, Estonia	9th Baltic Sea Tourism Forum
8 November 2016	Stockholm, Sweden	18th Baltic Development Forum Summit
8–9 November 2016	Stockholm, Sweden	7th EUSBSR Annual Forum
23–24 November 2016	Riga, Latvia	2nd Baltic Maritime Spatial Planning (MSP) Forum
22–23 March 2017	St. Petersburg, Russia	18th International Environmental Forum "Baltic Sea Day"
13 June 2017	Berlin, Germany	19th Baltic Development Forum Summit
13–14 June 2017	Berlin, Germany	8th EUSBSR Annual Forum
20–22 September 2017	Potsdam, Germany	25th Baltic Sea States Sub-regional Co-operation (BSSSC) Conference
2–3 November 2017	Turku, Finland	10th Baltic Sea Tourism Forum
22–23 March 2018	St. Petersburg, Russia	19th International Environmental Forum "Baltic Sea Day"
4–5 June 2018	Tallinn, Estonia	9th EUSBSR Annual Forum
10–12 September 2018	Gdánsk, Poland	26th Baltic Sea States Sub-regional Co-operation (BSSSC) Conference
14–15 November 2018	Riga, Latvia	11th Baltic Sea Tourism Forum
21–22 March 2019	St. Petersburg, Russia	20th International Environmental Forum "Baltic Sea Day"
12–13 June 2019	Gdánsk, Poland	10th EUSBSR Annual Forum
18–20 September 2019	Klaipeda, Lithuania	27th Baltic Sea States Sub-regional Co-operation (BSSSC) Conference
13 November 2019	Brussels, Belgium	12th Baltic Sea Tourism Forum
19–21 November 2019	Riga, Latvia	3rd Baltic Maritime Spatial Planning (MSP) Forum
16–18 September 2020	St. Petersburg, Russia	21st International Environmental Forum "Baltic Sea Day"
29 September–1 October 2020	Online-Conference	28th Baltic Sea States Sub-regional Co-operation (BSSSC) Conference
19–20 October 2020	Turku, Finland (Online)	11th EUSBSR Annual Forum
11–13 November 2020	Palanga, Lithuania	13th Baltic Sea Tourism Forum

2020b). Further events with a broad strategic focus are the Baltic MSP Forum, the BSSSC Annual Conference, and the International Environmental Forum "Baltic Sea Day." Beyond these, there are many other events relating to specific sectors (e.g. tourism, fisheries, energy), which may further underpin the supposed advanced status of the BSR in realising Blue Growth.

11.3.2 "Baltic Blues"? An Assemblage Perspective

Translating the presented descriptions into conceptual language, we argue that the interplay of actors, networks, discourses, strategies, action plans, and events brings about an agentic-centred, fluctuating, and contested constellation or, in other words, a Blue Economy assemblage. The strategic orientations show a certain degree of variability, fragmentation, and contradiction, enforced by a Blue Economy agenda that is clearly informed by economic growth narratives. From an assemblage perspective, the current status of the BSR has arisen from a complex web of "investment institution trajectories, with distinctive economy-environment relations" (Winder & Le Heron 2017, p. 7), which in turn will entail new entanglements and future trajectories. Strategically governing such trajectories in the direction desired is in fact virtually impossible, given that policy frameworks will be far from sufficiently preparing for both intended and unintended encounters of human and non-human entities, and their unforeseeable consequences, in marine and coastal spaces. Nevertheless, we may provide some entry points for thinking of the emerging Blue Economy in the BSR through the lens of assemblage theory.

The first entry point refers to the composition and articulation of Blue Economy policies. Processes of policy-building usually rely on the involvement of stakeholders in order to develop (or pretend?) a holistic, democratic, inclusive, and transparent governance approach. But who are stakeholders? Who identifies and defines them? What is the best way to foster dialogue? Deeper analysis of Blue Economy policies in the BSR implies further questions on the ontology of their becoming, especially when considering the role of language and discourse. It is striking that the term "blue" is used in different ways to frame maritime and coastal economies. Beside Blue Economy and Blue Growth, we found other notions appearing in strategic documents such as blue bioeconomy, blue biotechnology, and blue-green economy. These diverse applications of the "Blue" demonstrate certain arbitrariness, while each notion constitutes a different assemblage including and excluding specific sectors, actors, practices, and relations. It must be questioned whether such terminological inconsistency proves a viable way for sustainable BSR development. The multiplicity of policies, programmes, and action plans may further express what Winder and Le Heron (2017, p. 11) refer to as "heterogeneous investment institutional initiatives of quasi collective shape." The strong influence of the EU raises additional questions on the multi-scalar nature of policy-building affecting marine and coastal spaces. In this vein, Garland et al. (2019) point to the emergence of "scale mismatches" and their rather neglected role in Blue Economy discourses. Against this backdrop, the translation of the EU-driven Blue Growth agenda to the BSR and its national adoption by the Baltic Sea states should be discussed more intensely.

This leads to the second entry point, namely, the contentious economic imperative that is related with Blue Growth. As already mentioned, the promise behind this policy narrative is to achieve smart, sustainable, and inclusive growth from marine and coastal resources. This ambition seems reasonable, as it suggests that economic, environmental, and social dimensions are equally considered. However, by defining four "high potential areas"—shipping, blue bioeconomy, coastal and maritime tourism, environmental and monitoring technology—Blue Growth sets clear priorities, while at the same time de-prioritising, if not excluding, other sectors. For example, fisheries are not explicitly addressed, which inevitably raises the question of how inclusive Blue Growth actually is? Moreover, what does it tell us about environmental issues, when maritime tourism (including cruise-shipping) is given high priority? From this viewpoint, the deliberate linking of attributes such as "smart," "sustainable," and "inclusive" with the "growth" metaphor may be considered a strategic move to achieve broad legitimation for Blue Growth. Therefore, this example indicates how single policies resulting from the interaction of heterogeneous components, such as actors, organisations, documents, or language, may create an own agency, while playing material and symbolic roles, in a Blue Economy assemblage. Blue Growth further demonstrates the obvious enforcement of economic interests over social and environmental issues, which in turn reflects specific constellations of power within multi-scalar governance assemblages (including EU and BSR marine policy settings).

This directly leads to a third entry point in that we take a closer look at the construction of economy-environment relations. This is probably one of the most contested issues in developing a Blue Economy, as actors seek either to rework or to avoid reworking such relations (Winder & Le Heron, 2017). In this regard, the "Baltic Sea Day" taking place annually in St. Petersburg, Russia, is a worrying example. Conceived as a key platform for environmental dialogue and supported by HELCOM, the "Baltic Sea Day" aims at "sharing experiences (…) towards the ambitious goals for the conservation and protection of the Baltic Sea environment" (HELCOM, 2020a). These goals, however, are strongly thwarted by the fact that the event is sponsored and organised by the Russian state company "Mineral." Moreover, this company is designated as the national coordinator of HELCOM activities in Russia. Keeping in mind that HELCOM's future vision is "a healthy Baltic Sea environment with diverse biological components (…)" (HELCOM, 2020b), there are no further explanations regarding credibility and environmental justice required. We argue that such entanglements of economic and environmental interests are an immanent and recurrent element of each Blue Economy with both stabilising and destabilising effects. This point of departure offers a lot of possibilities for rethinking economy-environment relations in the making of a Blue Economy, while also recognising political and ethical challenges.

11.4 Critical Outlook

From an assemblage perspective, the Blue Economy agenda for the BSR can be conceived as a fluctuating and fragmented project based on multi-scalar, multi-stakeholder, and multi-sited processes. The diversity of policies reflects the heterogeneity of components that are assembled into competing or shared sets of political arrangements for the purpose of governing the BSR. However, the outcome of the underlying negotiations—i.e. visions, objectives, action plans, etc.—inevitably causes prioritisation and de-prioritisation or, more pointedly, inclusion and exclusion of specific sectors and stakeholders. This encourages critical engagement with the meaning of sustainability in the regional Blue Economy discourse. Several authors clearly criticise the economic orientation of the EUSBSR and the related *Sustainable* Blue Growth Agenda (e.g. Ertör & Hadjimichael, 2020; Schultz-Zehden et al., 2019). As Silver et al. (2015, p. 153) put it, "the European Union has perhaps most overtly tied the term Blue Economy to capitalization and accumulation," while Garland et al. (2019, p. 10) state that "in defining the boundaries of the BE, the EU framed this clear regional economic paradigm within the boundaries of [an] environmental sustainability." In this light, we argue that assemblage thinking offers a promising lens for both uncovering the fragile relationalities in specific Blue Economy settings and questioning the inherent imaginaries of sustainability that often appear taken-for-granted.

Thinking about a sustainable Blue Economy agenda that only works when the social, the economic, and the environment (whatever these are) do "assemble" implies an urgent need to "invest" in the social dimension. Further research should steer in this direction, as many contributions tend to be biased to the economy-environment nexus. By contrast, the situation of local coastal communities that still depend on traditional uses of marine resources for their day-to-day livelihoods is not sufficiently addressed (Schultz-Zehden et al., 2019). While traditional sectors (e.g. small-scale fisheries) do not play a major role within EU Blue Economy policies, at least with respect to the BSR, the prospects for local communities have to be studied in more detail. It is important to understand how these communities are affected by defined Blue Growth sectors such as "blue bioeconomy" or maritime and coastal tourism, and how far the "locals" could *really* participate in the process of policy-building. Answering these questions may provide a promising account for connecting society more effectively to the ocean, coasts, and seas and to realise a more sustainable and inclusive Blue Economy across economic, ecological, social and cultural scales ((McKinley, 2023), this volume)). We are convinced this will help to overcome the "lack of an integration of sustainability and social justice principles" (Garland et al., 2019, p. 11) within emerging Blue Economy settings.

References

Anderson, B., Kearnes, M., McFarlane, C., & Swanton, D. (2012). On assemblages and geography. *Dialogues in Human Geography, 2*(2), 171–189.

Baker, T., & McGuirk, P. (2017). Assemblage thinking as methodology: Commitments and practices for critical policy research. *Territory, Politics, Governance, 5*(4), 425–442.

Briassoulis, H. (2019). Governance as multiplicity: The Assemblage Thinking perspective. *Policy Sciences, 52*(3), 419–450.

Duffy, P., & Stojanovic, T. (2018). The potential for Assemblage thinking in population geography: Assembling population, space, and place. *Population, Space Place, 24*(3), e2097.

ECORYS, Deltares, & Oceanic Development, (2012): *Blue Growth—Scenarios and Drivers for Sustainable Growth from the Oceans, Seas and Coasts. Final Report*. Brussels and Rotterdam. Retrieved 20 July 2020, from https://webgate.ec.europa.eu/maritimeforum/system/files/Blue%20Growth%20Final%20Report%2013092012.pdf

Ertör, I., & Hadjimichael, M. (2020). Editorial: Blue degrowth and the politics of the sea: Rethinking the Blue Economy. *Sustainability Science, 15*(1), 1–10.

European Commission (EC). (2014). *A Sustainable Blue Growth Agenda for the Baltic Sea Region. Commission Staff Working Document*. Brussels: European Commission (EC). Retrieved 20 July 2020, from https://ec.europa.eu/maritimeaffairs/sites/maritimeaffairs/files/docs/body/swd-2014-167_en.pdf.pdf

European Commission (EC). (2019). *Interreg Baltic Sea Region 2021–2027. Orientation Paper*. Retrieved 27 August 2020, from https://www.interreg-baltic.eu/fileadmin/user_upload/Future_programme/Baltic_Sea_Region_Orientation_Paper.pdf

European Commission (EC). (2020a). *The EU Blue Economy Report 2020*. Luxembourg: Publications Office of the European Union.

European Commission (EC). (2020b). *EU Strategy for the Baltic Sea Region—Action Plan. Agreed by the NC Group*. Retrieved 18 August 2020, from https://www.balticsea-region-strategy.eu/news-room/news/591126-eusbsr-action-plan-approved-by-the-national-coordinators-group

European Union (EU). (2020a). *The Interreg Baltic Sea Programme. Programme Fact Sheet*. Retrieved 27 August 2020, from https://www.interreg-baltic.eu/about-the-programme.html

European Union (EU). (2020b). *11th Annual Forum of the EUSBSR*. Retrieved 28 August 2020, from http://www.annualforum2020.eu/en

Fairbanks, L., Campbell, L.M., Boucquey, N., & St. Martin, K. (2018). Assembling enclosure: Reading Marine Spatial Planning for alternatives. *Annals of the American Association of Geographers, 108*(1), 144–161.

Garland, M., Axon, S., Graziano, M., Morrisey, J., & Heidkamp, C.P. (2019). The Blue Economy: Identifying geographic concepts and sensitivities. *Geography Compass, 13*(7), e12445.

Gänzle, S., & Kern, K. (2016). The European Union strategy for the Baltic Sea region. In Gänzle, S., & Kern, K. (Eds.), *A 'Macro-regional' Europe in the Making. Theoretical Approaches and Empirical Evidence* (pp. 123–144). London: Palgrave Macmillan.

Hassler, B., Gee, K., Gilek, M., Luttmann, A., Morf, A., Saunders, F., Stalmokaitė, S., Strand, H., & Zaucha, J. (2019). *New generation* EU directives, sustainability, and the role of transnational coordination in Baltic Sea maritime spatial planning. *Ocean & Coastal Management, 169*, 254–263.

Hegland, T.J., Raakjær, J., & van Tatenhove, J. (2015). Implementing ecosystem-based marine management as a process of regionalisation: Some lessons from the Baltic Sea. *Ocean & Coastal Management, 117*, 14–22.

Helsinki Commission (HELCOM). (2018). *Implementation of the Baltic Sea Action Plan 2018: Three Years Left to Reach Good Environmental Status. Background Document to the 2018 HELCOM Ministerial Meeting 2018, Brussels.* Retrieved 21 August 2020, from https://helcom.fi/wp-content/uploads/2019/06/Implementation-of-the-BSAP-2018.pdf

Helsinki Commission (HELCOM). (2020a). Official Website: Events. Retrieved 09 September 2020, from https://helcom.fi/helcom-at-work/events/events-2020/baltic-sea-day-2020/

Helsinki Commission (HELCOM). (2020b). Official Website: About us. Retrieved 09 September 2020, from https://helcom.fi/about-us/

Kristensen, A.R., Lopdrup-Hjorth, T., & Sorensen, B.M. (2014). Gilles Deleuze (1925–1995). In Helin, J., Hernes, T., Hjorth, D., & Holt, R. (Eds.), *The Oxford Handbook of Process, Philosophy and Organization Studies* (pp. 499–514). Oxford: Oxford University Press.

Marks, J. (2005). Thought. In Parr, A. (Ed.), *The Deleuze Dictionary*, revised ed. (pp. 284–286). Edinburgh: Edinburgh University Press.

Mattissek, A., & Wiertz, T. (2014). Materialität und Macht im Spiegel der Assemblage-Theorie: Erkundungen am Beispiel der Waldpolitik in Thailand. *Geographica Helvetica, 69*, 157–169.

McFarlane, C., & Anderson, B. (2011). Thinking with assemblage. *Area*, *43*(2), 162–164.

McKinley, E. (2023). A Blue Economy for who? Linking marine social sciences with Blue Economy discourse. In Heidkamp, C.P., Morrissey, J., & Germond-Duret, C. (Eds.), *Blue Economy: People and Regions in Transitions*. London & New York: Routledge.

Morf, A., Moodie, J., Gee, K., Giacometti, A., Kull, M., Piwowarczyk, J., Schiele, K., Zaucha, J., Kellecioglu, I., Luttmann, A., & Strand, H. (2019). Towards sustainability of marine governance: Challenges and enablers for stakeholder integration in transboundary marine spatial planning in the Baltic Sea. *Ocean & Coastal Management, 177*, 200–212.

Müller, M., & Schurr, C. (2016). Assemblage thinking and actor-network theory: Conjunctions, disjunctions, cross-fertilisations. *Transactions of the Institute of British Geographers, 41*(3), 217–229.

Ollikainen, M., Hasler, B., Elofsson, K., Iho, A., Andersen, H.E., Czajkowski, M., & Peterson, K. (2019). Toward the Baltic Sea socioeconomic action plan. *Ambio, 48*, 1377–1388.

Savage, G.C. (2019). What is policy assemblage? *Territory, Politics, Governance, 8*(3), 319–335.

Schultz-Zehden, A., Weig, B., & Lukic, I. (2019). Maritime spatial planning and the EU's Blue Growth policy: Past, present and future perspectives. In Zaucha, J., & Gee, K. (Eds.), *Maritime Spatial Planning* (pp. 121–149). Cham: Palgrave Macmillan.

Silver, J.J., Gray, N.J., Campbell L.M., Fairbanks, L.W., & Gruby, R.L. (2015). Blue Economy and competing discourses in international oceans governance. *Journal of Environment and Development, 24*(2), 1–26.

Söderström, S., & Kern, K. (2017). The ecosystem approach to management in marine environmental governance: Institutional interplay in the Baltic Sea region. *Environmental Policy and Governance, 27*(6), 619–631.

s.Pro. (2017). *Towards an Implementation Strategy for the Baltic Blue Growth Agenda. Commissioned by the European Commission.* Luxembourg: Publication Office of the European Union.

SUBMARINER Network for Blue Growth EEIG. (2015). *SUBMARINER Roadmap. Towards a Blue-green Economy in the Baltic Sea Region*, 2nd ed. Gdańsk.

SUBMARINER Network for Blue Growth EEIG. (2018). *SUBMARINER Flyer.* Retrieved 19 August 2020, from https://www.submariner-network.eu/images/Submariner_flyer-2.pdf

Westholm, A. (2018). Appropriate scale and level in marine spatial planning—Management perspectives in the Baltic Sea. *Marine Policy, 98*, 264–270.

Winder, G., & Le Heron, R. (2017). Assembling a Blue Economy moment? Geographic engagement with globalizing biological-economic relations in multi-use marine environments. *Dialogues in Human Geography*, 7(1), 3–26.

World Wildlife Fund (WWF). (2015). *WWF Report: All Hands on Deck—Setting Course Towards a Sustainable Blue Economy.* Retrieved 20 August 2020, from https://wwfeu.awsassets.panda.org/downloads/15_6802_all_hands_on_deck_lr_151008.pdf

PART 3

Blue Economy—Futures

Blue Economies in Transitions

12

DECIDING PORT FUTURES

Ports of Auckland, Marine Spatial Planning, and Contested Ethics in Blue Economy Plan Making

Marie Aschenbrenner and Gordon M. Winder

12.1 Introduction

Public debate has swirled in Auckland, Aotearoa New Zealand (ANZ), over the future of the city's port as expectations concerning the performance of the city's Waitematā Harbour (the Harbour) and Hauraki Gulf Tīkapa Moana (the Gulf) change. The port's presence on Auckland's inner-city waterfront and its continued use of shipping lanes in the Waitematā Harbour and Gulf are being increasingly challenged. Diverse actors raise the following questions: should port activities be relocated, freeing the waterfront and channel approaches for other, more profitable, more environmentally compatible, and less industrial uses; what is the "right" decision concerning the port's future; and what constitutes "good" decision-making on this contested issue?

Analysing how decisions are made and negotiated in this issue, with a focus on governance, helps to gain further understanding of coastal transitions—here energised by new expectations arising from Blue Economy (BE) accounting and marine spatial planning (MSP). In 2013–2016, MSP was introduced in the Gulf in an effort to "implement an ecosystem-based approach to the marine environment" (Peart, 2019, p. 1). From 2014, BE received increasing attention due to a national science initiative on Sustainable Seas (Sustainable Seas, 2020). In this research, the economic value creation potential of the Harbour and Gulf are highlighted due to their "prominence in the national marine economy" (Envirostrat, 2019, p. 23). Local government and business are increasingly interested in investing in sea-dependent and related value chains (Envirostrat, 2019; Winder & Le Heron, 2017). Efforts to generate greater value from the Gulf's BE while planning for healthier marine ecosystems are inevitably entangled with further expectations, most notably Māori rights and entitlements. Thus, in Auckland, BE and MSP must be seen as part of rapidly changing and competing

DOI: 10.4324/9781003280248-16

agendas. In a buzz of excited engagement, economic-environment projects and relations are being reimagined, reworked, and renegotiated. The way bargaining and decision-making takes place—leading to a transformation of the overall socio-natural assemblage and, from a political ecology point of view, to certain beneficiaries and losers—is an important aspect of the human dimensions in ocean governance and management, and thus for the marine social sciences (Bennett, 2019b; Heidkamp & Morrissey, 2019).

This chapter analyses the role of ethics in decision-making. Detecting ethical claims-making and a vocabulary of ethics in the contested port issue, the question of an "ethicization of the conflict" is raised, and how this connects to framing and decision-making in BE projects. The analysis draws on 27 semi-structured and narrative interviews conducted with actors in and around

TABLE 12.1 Expert interviews conducted during the research period, 2018–2020

Interview		*Fields of expertise**	*Research phase***
1	Key Informant 1	SCTTTP***, Social Sciences	1
2	Key Informant 2	SCTTTP, Planning Consultant	1
3	Key Informant 3	SCTTTP, Auckland Council	1
4	Key Informant 4	Planning Consultant	1
5	Key Informant 5	SCTTTP, Environmental Conservation	1
6	Key Informant 6	SCTTTP, Social Sciences	1
6	Key Informant 7	SCTTTP, Social Sciences	1
7	Key Informant 8	Sea Change Tai Timu Tai Pari (SCTTTP), Waikato Regional Council	2
8	Key Informant 9	SCTTTP, Marine Biology and GIS	2
9	Key Informant 10	SCTTTP, Mātauranga Māori	2
10	Key Informant 11	Auckland Council, SCTTTP	2
11	Key Informant 12	SCTTTP, Department of Conservation	2
12	Key Informant 13	Media, Environmental Conservation	2
13	Key Informant 14	SCTTTP, Hauraki Gulf Forum	2
14	Key Informant 15	SCTTTP, Mātauranga Māori	2
15	Key Informant 16	Urban Development, SCTTTP, Mātauranga Māori	2
16	Key Informant 17	Waterfront Development, Auckland Council	2
17	Key Informant 18	SCTTTP, Mātauranga Māori	2
18	Key Informant 19	SCTTTP, Recreational Fisheries	2
19	Key Informant 20	SCTTTP, Hauraki Gulf Forum	2
20	Key Informant 21	Department of Conservation, SCTTTP	3
20	Key Informant 22	Ministry for Primary Industries, SCTTTP	3
21	Key Informant 23	Cruise Ship Industry and Port Issues	3
22	Key Informant 24	Ports of Auckland Ltd., SCTTTP	3
23	Key Informant 25	Protests in the Context of POAL extensions	3
24	Key Informant 26	Protests in the Context of POAL extensions	3
25	Key Informant 27	Planning consultant, Blue Economy, SCTTTP	2, 3

**Listed according to the key focus in the interview.*

***Research phases: October–December 2018 (1), February–April 2019 (2), January–July 2020 (3).*

****SCTTTP stands for Sea Change Tai Timu Tai Pari, the marine spatial planning process in the Gulf.*

the Gulf between late 2018 and July 2020 (Table 12.1). Interview partners were experts in an MSP process (Sea Change Tai Timu Tai Pari) and for the Gulf's overall BE. Eight of them were directly involved in port-related issues and struggles through their work for respective organisations (Key Informants 11, 16, 17, 23, 24, 25, 26, 27). The interviews focused on different fields of expertise and topics, namely, MSP, BE, marine conservation, port-related issues, and struggles as well as experts' perceptions and imaginations of the Gulf and Harbour. This chapter also makes use of analysis of reports, associated material, and media coverage. This research is supported by a DFG (German Research Foundation) grant and is a project within the DFG-funded Research group "Urban Ethics." A special focus is on the ethics of "scientific knowledge and expertise [produced in the process] to increase moral authority, credibility, or legitimacy" (Bennett, 2019a, p. 72), techniques of governing and the creation (and challenging) of appropriate subject positions, both having their part in urban ethical negotiations (Dürr et al., 2019; Foucault, 1987).

This chapter first provides a review of developments in port industry, urban port and waterfront areas, and their analysis in geography literature. Connections are made between Political Ecology and an Urban Ethics viewpoint. The following sections set current tensions around the Gulf, port of Auckland, and the port's narratives and perspectives in context. On that basis, struggle over BE decision-making over Auckland port futures is explored with a special focus on the role of ethical claims making. The main findings indicate the important and multi-layered role of ethics in BE thinking, legitimising specific rationalities, resource uses, subject positions, and sorting out winners and losers.

12.2 Port Relocation and Waterfront Development

Port relocation with subsequent redevelopment of urban waterfronts is not a topic new to the urban development, planning or human geography literatures (see Ng & Ducruet, 2014 for a recent review of port geography literature). At the end of the last century, containerisation and mechanisation facilitated growing port productivity as increasing vessel sizes combined with improved intermodal freight handling, transforming seaports worldwide (Frémont, 2013; Olivier & Slack, 2006). These changes had a "profound effect on port structure, port operation [and] traditional port functions" (Hayuth, 1981, p. 160) as well as the "long-standing spatial and functional ties between ports and cities" (Hayuth, 1982, p. 219).

As privatisation and commercialisation proceeded, transnational corporations and public-private partnerships ousted public authorities from port-operating and owning (Brooks, 2004; Heaver, 2002). "Industrial and shipping agencies have reformulated their positions with non-port-related interests [such as] in the struggles to determine primary land uses on the waterfront" (Bunce & Desfor, 2007, p. 252). New "patterns of economic activities and [..] technological developments [were regarded in the literature as] primary forces" (Bunce & Desfor, 2007, p. 252) in the closing down and outward migration of terminals and port-related industry, leaving behind vast inner-city areas often treated as "urban areas of discard" until developed in waterfront renewal projects (Hilling, 1988; Hoyle, 1989; Slack, 1980, 1989; Suykens, 1989). More recent work on waterfronts has criticised these narratives and models for failing to "recognise[e] the complexity and dynamics of waterfront regeneration" (Brownill, 2013, p. 50; Desfor et al., 2011). Instead, authors of a 2007 special issue of *Cities* understand urban waterfronts as "contested arenas" with numerous actors involved, embedded in different social and ecological processes, networks of power, and connected to contradictory meanings, identities, desires, urban imaginaries, discourses and modes of governance (Bunce & Desfor, 2007; Dodman, 2007; Hagerman, 2007; Kear, 2007; Laidley, 2007; O'Callaghan & Linehan, 2007; Wakefield, 2007).

While covering very different cities, contexts, and projects, much research still focuses on transformation after a port's waterfront footprint is reduced through relocation or truncated activity (for an exception, see, e.g., Ou & Ma, 2017). In Auckland, the usual changes in port operations and structures can be observed, but, interestingly, the main port facilities remain located at the waterfront while debate over the port's future continues. Especially in a time when BE accounting receives growing attention in urban contexts (UN-Habitat, 2018), contestation over port relocation presents an intriguing research topic. Potentially, classic port-related activities, "waterfront restoration," and upcoming services such as carbon offsetting are all BE sectors and meet here. They connect to different local and global discourses, urban imaginaries, and developments but collide with a "classic" redevelopment process. In this context, no single way of decision-making, accounting, and argumentation is accepted or normative. Instead, the grounds on which the future of the port (and waterfront) are decided are fiercely contested. This text aims not at working out the "right" decision in this struggle but at understanding the dynamics and grounds on which actors claim "right" (or efficient/effective/etc.) decision-making. How do actors legitimise their positions, how do they build credibility, and contest others in a field where planning and decision-making practices seem subjective, open, and in the making?

Adopting a political ecology approach, this chapter frames the Harbour and Gulf as a "multi-faceted conflict zone" in which different groups negotiate access to natural resources (Robbins, 2012). This helps to question the "rationalist paradigm" (Flannery & Ellis, 2016, p. 123) of planning and decision-making practices, and prevalent scientific and moral discourses of appropriate and legitimate resource use (Boucquey, 2017). Geographers researching MSP, fisheries management, and marine conservation have identified how certain techniques of government—participatory processes, peer review, and expert panels—are used to claim neutrality for the apparatus of decision-making, thus depoliticising decisions while marginalising opposition and obscuring winners and losers (Aschenbrenner & Winder, 2019; Bennett, 2019a; Ellis & Flannery, 2016; Tafon, 2018).

However, this chapter goes one step further by conceptualising the debates around the relocation of Auckland's port as a "field of Urban Ethics" (Dürr et al., 2019). To date, ethics—as discursive fields in which different views, interests, problematisations, guiding principles, and ideals of the good are being negotiated—have received little attention in this context (Fischer, 2020). While technological, socio-economic, and political logics are all present in Auckland, they appear to be embedded in a field where "good" political behaviour, "right" economic assessment, or "sustainable" technologies and practices are negotiated. Ethical claims seem to be "the oil in the machinery of decision making," the means with which actors interact, negotiate, and claim certain practices, knowledge, and urban futures. Here, the way in which actors refer to moral imperatives or principles are in the centre of the analysis. Due to space constraints, the focus is on power holders rather than urban dwellers', port workers' and others' "ordinary ethics" which are essential aspects for further research (Lambek, 2010). In turn, this approach allows a focus on how ethical problematisations and claims-making function to (de)legitimise positions and speakers in debates or to create credibility and support for apparatuses of decision-making.

12.3 Context: Harbour and Gulf as Multifaceted Conflict Zone

When seen through a political ecology lens, the Harbour and Gulf are not a "natural," politically inert system. In the historical transformation of the Gulf and Harbour, one finds "myriad 'social-natural' productions" (Bunce & Desfor, 2007) each having its own set of narratives and socio-natural relations. Today, different knowledge, ideas, and agendas on how socio-natural relations of the Gulf, Harbour, and coast should be used and managed come together and overlap (Le Heron et al., 2019).

On a formalised governance level, several local institutions (and their rules and agendas) are involved, together with multiple national authorities. In 2000, national government established both a marine park with authority for the Gulf and its catchments, and the Hauraki Gulf Forum, with its own agenda of integrating management of the area among different governmental agencies and *tangata whenua* ('people of the land') (Peart, 2017).

The rights of Māori as tangata whenua, Māori knowledge systems, interests, and associations to land and sea have rarely been recognised on a government level (Le Heron et al., 2019). A settlement process for grievances related to government failures to adhere to the Treaty of Waitangi, an agreement made in 1840 between representatives of the British crown and Māori iwi/hapū ("tribes"/"subtribes"), remains incomplete. While fisheries claims were settled in the 1990s, multiple claims to the foreshore and seabed of the area remain unsettled (Hauraki Gulf Forum et al., 2016). Recent government initiatives to ensure Māori rights and participation and the representation of Māori ontology in (environmental) management and governance have been made but are often met with political resistance (Le Heron et al., 2019). While iwi/hapū share common themes, nuances of opinion and different views persist.

In addition, Sea Change, a non-statutory, collaborative MSP exercise, aimed at a consensual agreement between Treaty partners, governmental agencies, and other actors in and around the Gulf for its better use and environmental protection—an attempt to reimagine and rework socio-natural relations. The Gulf's environmental state is reported as degraded (Hauraki Gulf Forum, 2020). MSP planning followed an ecosystem approach while integrating Māori worldviews and values (Hauraki Gulf Forum et al., 2016). The process revealed multifaceted tensions between different socio-natural relations, discourses, and ethical perspectives (key informants 1–3, 5–16, 18–22, 24, 27).

Controversies and competing agendas are evident in tensions around specific projects and initiatives, whether marina developments (SKP, 2017) or wastewater management (Acosta García et al., 2020). Thus, the "port case" is only one of many foci where questions around urban, socio-natural relations are contested, negotiated and (re)produced.

12.4 The Port in the Hauraki Gulf Tīkapa Moana

The port has long been a major player in the Gulf. It was imposed at what is now Auckland's isthmus on Māori traditional land and waters. Back then Auckland was "rather a port city than a city with a port" (Winder, 2006, p. 56): the city grew around the wharves. Until 1988, over 122 hectares of land were reclaimed for the port "dramatically expanding the city's high value commercial real estate" (Winder, 2006, p. 55) and heavily influencing the material forms of the Harbour (Bunce & Desfor, 2007). The waterfront was assembled as a site of functioning port operations, according to principles of efficiency, use orientation, and exclusion for reasons of safety.

In 1988, Auckland Harbour Board (AHB) was turned into Ports of Auckland Limited (POAL) with ANZ's government aiming for a change towards greater commercialisation, efficiency, and productivity—"every port [was] to operate as successful business" (Pyvis & Tull, 2017; New Zealand Government, 2014, p. 8). POAL became 100% council owned in 2010, when, in a further reorganisation, the government amalgamated all of the local councils in Auckland region to form one unitary governing body. While public in function (it returns a dividend to the Auckland Council), POAL is not directly council controlled. An independent board of directors appointed by but not comprising members of Auckland Council works towards maximising the values of Auckland Council as shareholder and reports back to it. POAL has a major economic role in the city (Lewis & Murphy, 2015; POAL, 2022). With their control of vital sections of the city's extensive waterfront, POAL and AHB, their engineers and logistics experts, have long dominated the landscapes, seascapes, and environments of the harbour and Gulf (Winder, 2006).

12.5 Challenges to POAL

In the mid-2010s, POAL found its established, legitimated, and approved port management logic and position being challenged by alternative assessments of land development on Auckland's waterfront. In part, this owed to changes in the built environment, public discourses, and urban imaginary of the waterfront (Lewis & Murphy, 2015). Beginning with hosted America's Cup events in 2000 and 2003, redundant port space was remediated and developed through public-private investments, with the Viaduct Harbour as a first, "powerfully evocative position" (Murphy, 2008, p. 2537) in an ongoing redevelopment process. In the City's new rhetoric, a "liveable" urban waterfront environment is imagined—and supported by public opinion (Key Informants 17, 25, 26, NZ Herald, 2011). The waterfront should be a dense, diverse, vital, and mixed-use space where sea views, Pacific, and maritime cultural heritage come together to build a competitive asset for quality of life and to attract visitors, investment, and economic development (Auckland Council, 2012; Panuku Development, 2014).

In 2015, contradictions between the new imaginary and port use became obvious when POAL's proposal to further extend Bledisloe Wharf, a multi-cargo wharf close to the publicly accessible Queens Wharf, met with defeat. Urban Auckland, a group of architects and urban designers, stopped POAL's plan by High Court decision. Another network, Stop Stealing Our Harbour (now Waterfront 2029), emerged, joining business and community actors to protest against reclamation works and advocate for a relocation of Auckland's port. Both groups connect to the global discourse on waterfront redevelopments. They challenged POAL on "ethical terms": calling the company's public engagement "poor," its behaviour "unethical" and "arrogant" (Key Informant 25), and criticising POAL's role and decision-making power as "illegitimate" (Key Informants 25, 26).

12.6 POAL's Perspective and Narratives

In the last two decades, POAL's performance reports reveal changed perspectives and narratives. Unsurprisingly, it boasts that it is responding to local and global expectations by adapting to global sea trade developments, introducing further port mechanisation, containerisation and automation, and accommodating larger vessels through extensive harbour dredging (Pyvis & Tull, 2017). It reports its own performance using narratives and metrics of trade, port efficiency, and capacity (Table 12.2) as well as its international connectedness. In 2010, POAL owned 76 hectares of land on the city's waterfront and managed numerous assets and operations, from port shares in Whangarei and at Marsden Point to logistics and other (port) services' companies.

While POAL's core performance indicators have remained the same, its narrative has changed. Already in 2010, it reported significant reductions to its carbon footprint. It now aims for zero emissions by 2040. Its purchase of the world's first full-size electric ship-handling tug and installation of the world's largest soil-based vertical garden on a car-handling building can be interpreted as further signs of a new narrative of (ecological) sustainability and responsibility (POAL, 2020a, 2020b). POAL also boasted diverse public interaction opportunities

TABLE 12.2 Performance Indicators for the Ports of Auckland, 2006–2019 (POAL, 2009-2019)*

Year	*TEU number*	*Cargo m tonnes*	*Vehicles number*	*Passenger number*	*Cruise ship calls*	*Total ship calls*	*Revenue NZ$m*	*Earnings EBIT NZ$m*	*Ordinary dividend NZ$m*
2006	688,077	4.6	185,159	Nd	49	1,736	159.3	62.3	31.8
2007	773,993	4.1	168,200	Nd	49	1,743	163.1	51.6	19.9
2008	840,993	3.6	173,373	Nd	70	1,738	169.4	52.8	22.8
2009	843,590	2.7	110,560	Nd	69	1,620	163.4	46.0	7.2
2010	867,368	2.8	129,811	Nd	62	1,426	165.0	52.0	22.0
2011	894,383	3.5	123,362	150,289	79	nd	175.4	54.7	18.0
2012	808,654	3.8	147,221	188,694	97	nd	175.4	49.0	20.1
2013	818,819	4.4	170,835	200,000	100	nd	186.6	55.4	29.5
2014	968,741	5.6	207,591	195,944	89	nd	219.9	84.5	66.6
2015	972,434	5.9	243,801	199,250	90	nd	217.3	77.8	41.7
2016	907,099	5.8	248,065	259,758	101	1,588	211.1	84.0	54.8
2017	952,331	6.5	297,383	230,571	100	nd	224.4	60.3	51.3
2018	973,722	6.8	297,678	272,060	108	nd	243.2	76.8	51.1
2019	939,680	6.5	255,252	330,088	127	nd	248.1	53.9	18.6

**Passenger numbers are not given for 2006–2010; total ship calls are not given for 2011–2015 and 2017–2019; EBIT is short for Earnings Before Interest and Tax.*

including a quarterly magazine, Community Reference Group, and events like SeePort Festival or Tāmaki Herenga Waka Festival taking place in otherwise not publicly accessible port areas. In 2013, POAL was involved in a collaborative process to save the population of Bryde's whales in the Gulf from deaths through ship collisions (Key Informant 24). POAL participated in the MSP. This further emphasised POAL's role and location in the Gulf and connected POAL to key actors, such as the Hauraki Gulf Forum. It constituted POAL as collaborative, engaged, self-governed, and self-responsible—as an "ethical" subject, a caring part of the city (Dürr et al., 2019). These are signs that POAL is responding to new ethical standards and expectations.

12.7 Ethics as "Oil in the Machinery of Decision Making"

When, in 2015, opposition to port development activities became obvious, then mayor Len Brown initiated the *Port Future Study*, which soon became a point of reference for discussions around the port's (and waterfront's) future. It assessed future port options and should establish a "reliable and authoritative [process] on which Aucklanders agree that their council can base decision making on" (Auckland Council, 2015a). The moral authority, credibility, and legitimacy of the study as decision-making technology were connected to ethical claims-making by (1) claiming an independent and transparent proceeding; (2) taking up global and local urban ethical discourses of participatory/collaborative decision-making by establishing mechanisms with two groups consisting, in varying parts, of members of "stakeholder organisations" and *mana whenua* (Māori iwi/hapū with authority over land and resources) (Baiocchi & Ganuza, 2017; Hauraki Gulf Forum et al., 2016, see also Key Informants 7, 14, 21, 22, 24; Le Heron et al., 2019); and (3) following consensual values in the production of scientific knowledge and expertise. As a result of the process, the diverse members of the Auckland community, represented by the stakeholders, would become constituted as "well-informed" citizens—as Council hoped (Auckland Council, 2015b; Collier & Lakoff, 2005).

While the assessment framework was established consensually, the private consultancy EY led the assessment as experts. Following the participants' propositions, EY considered options including an expansion of the ports' current area, a relocation of trade task, and waterfront redevelopment. It was by no means the first or last reassessment of Auckland's port location choices (EY, 2019; NZIER, 2017; Smellie, 2019), and, like its predecessors, it reaffirmed the current port location as best choice in the short to medium term on the basis of the logic of port management and economic analysis. In the long term, EY recommended to establish a port relocation option for freight outside the Gulf but to retain cruise functions on the Waitematā. An expansion beyond its existing position at the waterfront was not recommended because of mana whenua and community opposition. Urban Auckland's reactions to the process were positive. They acknowledged that it complied with their desires: it had an "environmental, social and cultural perspective" and discussion by stakeholders, "and this is what we were asking for. The mayor [...] bless him" (Key Informant 25).

However, as new construction and reclamation plans became public, both critique of Auckland Council and POAL, and public debates around the port's future intensified (NZ Herald, 2019c). Much of the ensuing struggle for a decision among the different (power-holding) actors occurred within the set "reporting framework." A peer review of the original study commissioned by the City of Auckland contended that measurement, indicators, and evidence were wrong. Claims about back room influence in the 2016 Port Study decision to retain the port where it is, arose (NZ Herald, 2019a, 2019b), and were countered by POAL with further publications and projects that demonstrated transparency and defended planning "ethics" (POAL, 2018, 2019).

In 2019, ANZ government hired EY and the consulting firm Sapere to conduct the *Economic Analysis of Upper North Island Supply Chain Scenarios* (MoT, 2020). The reports "recommended the managed closure of the [port] and the development of Northport as the best way forward" (Fox, 2019). Auckland's then-mayor Phil Goff challenged the study for being biased by the political interests of the leader of the NZ First political party and Deputy Prime Minister, Winston Peters, who sought to transfer port business to Northport: "Predetermination is wrong. They need to do the study properly and what we're doing is trying to help them produce a credible report" (NZ Herald, 2018).

POAL eventually "answered" with two subsequent reviews conducted by two consulting firms to demonstrate a "completely unbiased view" (POAL, 2019). POAL as well as the two reviewing companies claimed EY's study was "flawed," and the decision arbitrary (not-fact based) (POAL, 2019). Further, one of the reviewers called it "concerning" that EY's report proposed "a fundamental reorientation of Auckland's economic geography [while not identifying] the impacts on people, public infrastructure, housing, and businesses" (NZ Institute of Economic Research (NZIER), 2017). The ethics thus claimed functioned not only to gain further moral authority and credibility and delegitimise the agency and socio-natural imaginaries of others, but they also reinforced reports, accounting, participation, and expert knowledge as technologies to decide and claim urban "blue" economic-environmental relations as certain. Opposition moved in the same framing, challenging decisions based on faults in comparative accounting between the port's return and actual land value and negative values of environmental impacts (Key Informant 26).

Of course, debates around waterfront use involve further actors and fields. These cannot all be reported here. However, an analysis of *NZ Herald* articles from 2015 to 2020 in context of the port future study reveals only few actors, little mana whenua presence, and only a few opinions expressed in public debate. Lately, public debate seems to be changing. A very recent twist in the discourse challenges the ethics of report creation and consultancy. Media as well as Waterfront 2029 criticised expert planning, with the *NZ Herald* stating:

> "the new report is the twentysomethings in about the same number of years and basically it says: let's not do anything, and in case you don't agree, here's a really stupid option to waste your time and money deciding not to do [...] When you don't like one report, you get another. And another. You bury the whole thing under a deep pile of sludge" (NZ Herald, 2020).

12.8 Conclusion

As port activities and organisation have changed, there are different options opening up for urban waterfronts. Auckland's case shows how waterfronts are spaces where different urban blue economic-environment relations are contested and in the making. As political ecology studies of marine planning issues show, there is not one, "right" predetermined way to decide on and between waterfront futures. The analysis further indicates how apparently depoliticised planning techniques—notably BE accounting—proved subject to bargaining and politics and to be uncertain, unreliable, and easily challenged assessment practices.

Urban ethics appear as a discursive field in which a common decision-making basis in port issues (related to economic assessment logics) is established and legitimised. Ethical claims are advanced by all (power holding) participants in the struggle over Auckland port futures. Accusations flew concerning the ethics of assessment practices, (faulty) assumptions, and political influence. People, groups, narratives, imaginaries, even objects gained or lost moral authority, legitimacy, credibility, and agency in this field. This, as well as emerging hierarchies, exclusions, and marginalisation due to constituted ethical subjectivities and ethical claims-making shows an "ethical violence" (Butler, 2005) and suggests winners and losers to the process. Those who have the means, including the ethics, to connect to the process gain. They can claim credibility and legitimise their rationalities and imaginaries or delegitimise others. Those, such as mana whenua, with alternative framings, worldviews, and assessments, or missing the (financial) means, risk to be marginalised or must adjust to the framing. Port workers and marginalised communities were absent from port relocation debates. Thus, ethical framing has a role in legitimising a certain "appropriate" resource use, one which is closely connected to economic and port management logics. Further, ethics shapes political contest by reinforcing a hegemonic framing for decision-making and foreclosing debates about other metrics or the very purpose of the planning process (Flannery et al., 2018). Ethical framing of the decision-making process helps to transfer contest away from legitimised fields of expert knowledge, participation, and moral negotiations, but it does not remove contestation. Instead, the ethics of expertise, participation, and negotiation become foci for challenges. Ethics are crucial for opening up a new political field (Dürr et al., 2019; Mouffe, 2005). Aiming at a socially equitable and ecologically sustainable BE, involves taking transformative decisions and thus involves claims about normativity as well as ethics. To be inclusive, reflection is needed on the (urban ethical) discursive fields where decisions are constituted. In governance processes organised around BE accounting (and MSP) ethics must be worked out in advance of the deliberation process. Shared ethics, agreed in advance, establish, and sustain a common decision-making framework, logics, and technologies.

Auckland's inner-city waterfront is a unique context for examining BE accounting at work. This section of ANZ's coast is among the most expensive pieces of real estate in the country, yet it is dominated by a working port handling the largest single share of the country's import trade. It is for this reason that assessment of the values and issues of port relocation are such a special test of the capacity of BE accounting as a planning tool. In this case, its capacity was found to be wanting: the official planning process ran into competition among experts using separate metrics for assessing economic outcomes; the process was critiqued for its lack of attention to environmental outcomes; social outcomes were ignored; and Maori interests and voices were strangely silent in public discussions. Far from depoliticising coastal planning, BE practices could not, in this case, deliver a legitimising narrative that would drive through a re-imagined city waterfront in the highly conflicted terrain of Auckland's urban ethics.

Acknowledgements or Credits List

The theoretical development of and empirical research for this chapter took place within the Germany-based research group "Urban Ethics." The interdisciplinary research group was funded by DFG (German Research Foundation) grants from 2015 to 2021, with the project being part of the second funding period

(2018–2021). We are very grateful to the many people who directly and indirectly participated in the research of the project and made valuable contributions. Our thanks also go to the reviewer of the draft chapter who helped improve the text with their valuable comments.

References

Acosta García, R., Aschenbrenner, M., Dürr, E., & Winder, G. M. (2020). Re-imagining cities as ecosystems: environmental subject formation in Auckland and Mexico City. *Urban Research & Practice*, 1–16. https://doi.org/10.1080/17535069.2020.1811886

Aschenbrenner, M., & Winder, G. M. (2019). Planning for a sustainable marine future? Marine spatial planning in the German exclusive economic zone of the North Sea. *Applied Geography, 110*, 102050. https://doi.org/10.1016/j.apgeog.2019.102050

Auckland Council. (2012). *The Auckland Plan: the World's Most Liveable City.* Retrieved on 03 July 2020 from https://www.aucklandcouncil.govt.nz/plans-projects-policies-reports-bylaws/our-plans-strategies/Documents/auckland-plan-2012-full-document.pdf

Auckland Council. (2015a). *Port Future Study.* Retrieved on 09 July 2020 from http://www.portfuturestudy.co.nz/

Auckland Council. (2015b). *Port Future Study: About the Study.* Retrieved on 09 September 2020 from http://www.portfuturestudy.co.nz/about/

Baiocchi, G., & Ganuza, E. (2017). *Popular Democracy: The Paradox of Participation.* Stanford University Press. https://books.google.com.ua/books?id=HsT_jwEACAAJ

Bennett, N. J. (2019a). In political seas: engaging with political ecology in the ocean and coastal environment. *Coastal Management, 47*(1), 67–87. https://doi.org/10.1080/08920753.2019.1540905

Bennett, N. J. (2019b). Marine social science for the peopled seas. *Coastal Management, 47*(2), 244–252. https://doi.org/10.1080/08920753.2019.1564958

Boucquey, N. (2017). 'That's my livelihood, it's your fun': the conflicting moral economies of commercial and recreational fishing. *Journal of Rural Studies, 54*, 138–150. https://doi.org/10.1016/j.jrurstud.2017.06.018

Brooks, M. R. (2004). The governance structure of ports. *Review of Network Economics, 3*(2). https://doi.org/10.2202/1446-9022.1049

Brownill, S. (2013). Just add water: waterfront regeneration as a global phenomenon. In M. E. Leary & J. McCarthy (Eds.), *The Routledge Companion to Urban Regeneration* (pp. 45–55). Routledge. https://doi.org/10.4324/9780203108581.ch3

Bunce, S., & Desfor, G. (2007). Introduction to 'Political ecologies of urban waterfront transformations'. *Cities, 24*(4), 251–258. https://doi.org/10.1016/j.cities.2007.02.001

Butler, J. (2005). *Giving an Account on Oneself.* Fordham University Press.

Collier, S. J., & Lakoff, A. (2005). On regimes of living. In A. Ong & S. J. Collier (Eds.), *Global Assemblages* (pp. 22–39). Blackwell. https://doi.org/10.1002/9780470696569.ch2

Desfor, G., Laidley, J., Stevens, Q., & Schubert, D. (2011). *Transforming Urban Waterfronts: Fixity and Flow.* Routledge.

Dodman, D. (2007). Post-independence optimism and the legacy of waterfront redevelopment in Kingston, Jamaica. *Cities, 24*(4), 273–284. https://doi.org/10.1016/j.cities.2006.09.001

Dürr, E., Ege, M., Moser, J., Neumann, C. K., & Winder, G. M. (2019). Urban ethics: towards a research agenda on cities, ethics and normativity. *City, Culture and Society*, 100313. https://doi.org/10.1016/j.ccs.2019.100313

Ellis, G., & Flannery, W. (2016). Marine spatial planning: Cui bono?. *Planning Theory & Practice*, *17*(1), 122–128. https://doi.org/10.1080/14649357.2015.1131482

Envirostrat. (2019). *Transitioning to a Blue Economy: scoping and horizon scanning*. Retrieved on 27 August 2020 from https://www.sustainableseaschallenge.co.nz/assets/dms/Reports/Transitioning-to-a-blue-economy-Scoping-and-horizon-scanning/Envirostrat20Blue20Economy20Report202C20Dec20201920FINAL.pdf

EY. (2019). *Economic Analysis of Upper North Island Supply Chain Scenarios*. Retrieved on 10 July 2020 from https://www.transport.govt.nz/assets/Import/Uploads/Research/Documents/Cabinet-Papers/dd282d4e63/Attachment-1-Economic-Analysis-of-Upper-North-Island-Supply-Chain-Scenarios.pdf

Fischer, J.-M. (2020). *Urbane Ethiken und Umweltschutz: Ideale, Praktiken und Aushandlungen um die gute Stadt in Auckland, Aotearoa Neuseeland*. https://doi.org/10.14361/9783839450390

Flannery, W., & Ellis, G. (2016). Exploring the winners and losers of marine environmental governance. *Planning Theory & Practice*, *17*(1), 121–122. https://doi.org/10.1080/14649357.2015.1131482

Flannery, W., Healy, N., & Luna, M. (2018). Exclusion and non-participation in marine spatial planning. *Marine Policy*, *88*, 32–40. https://doi.org/10.1016/j.marpol.2017.11.001

Foucault, M. (1987). Zur Genealogie der Ethik: Ein Überblick über laufende Arbeiten. (Gespräch mit Hubert L. Dreyfus und Paul Rabinow). In H. L. Dreyfus & P. Rabinow (Eds.), *Michel Foucault: Jenseits von Strukturalismus und Hermeneutik* (pp. 268–292). Athenäum-Verl.

Fox, A. (2019). Major study says Ports of Auckland should close, Northport be developed. *NZ Herald*, October 3, 2020. https://www.nzherald.co.nz/business/news/article.cfm?c_id=3&objectid=12273168

Frémont, A. (2013). Containerization and intermodal transportation. In J.-P. Rodrigue, T. Notteboom, & J. Shaw (Eds.), *The SAGE handbook of transport studies* (pp. 47–62). SAGE Publications, Ltd.

Hagerman, C. (2007). Shaping neighborhoods and nature: urban political ecologies of urban waterfront transformations in Portland, Oregon. *Cities*, *24*(4), 285–297. https://doi.org/10.1016/j.cities.2006.12.003

Hauraki Gulf Forum. (2020). *State of our gulf 2020: state of the environment report*. Retrieved on 01 September 2020 from https://www.aucklandcouncil.govt.nz/about-auckland-council/how-auckland-council-works/harbour-forums/docsstateofgulf/state-gulf-full-report.pdf

Hauraki Gulf Forum, Ministry for Primary Industries, Department of Conservation, Waikato Regional Council, & Auckland Council. (2016). *Launch Copy: Tai Timu Tai Pari Sea Change—Hauraki Gulf Marine Spatial Plan*. Retrieved on 27 May 2019 from www.seachange.org.nz

Hayuth, Y. (1981). Containerization and the load center concept. *Economic Geography*, *57*(2), 160–176. https://doi.org/10.2307/144140

Hayuth, Y. (1982). The port-urban interface: an area in transition. *Area*, *14*(3), 219–224. http://www.jstor.org/stable/20001825

Heaver, T. D. (2002). The evolving roles of shipping lines in international logistics. *International Journal of Maritime Economics*, *4*(3), 210–230. https://doi.org/10.1057/palgrave.ijme.9100042

Heidkamp, C. P., & Morrissey, J. (2019). Introduction: sustainability in the coastal zone. In C. P. Heidkamp & J. Morrissey (Eds.), *Towards Coastal Resilience and Sustainability* (pp. 1–11). Routledge.

Hilling, D. (1988). Socio-economic change in the maritime quarter: the demise of sailortown. In B. S. Hoyle, D. A. Pinder, & M. S. Husain (Eds.), *Revitalizing the Waterfront: International Dimensions of Dockland Redevelopment* (pp. 20–37). John Wiley.

Hoyle, B. S. (1989). *The Port-City Interface: Trends, Problems and Examples*. Elsevier Ltd.
Kear, M. (2007). Spaces of transition spaces of tomorrow: making a sustainable future in Southeast False Creek, Vancouver. *Cities*, *24*(4), 324–334. https://doi.org/10.1016/j.cities.2007.01.005
Laidley, J. (2007). The ecosystem approach and the global imperative on Toronto's Central Waterfront. *Cities*, *24*(4), 259–272. https://doi.org/10.1016/j.cities.2006.11.005
Lambek, M. (2010). *Ordinary Ethics: Anthropology, Language, and Action*. Fordham University Press. http://ebookcentral.proquest.com/lib/ub-lmu/detail.action?docID=3239541
Le Heron, E., Logie, J., Allen, W., Le Heron, R., Blackett, P., Davies, K., … Hikuroa, D. (2019). Diversity, contestation, participation in Aotearoa New Zealand's multi-use/user marine spaces. *Marine Policy*, *106*, 103536. https://doi.org/10.1016/j.marpol.2019.103536
Lewis, N., & Murphy, L. (2015). Anchor organisations in Auckland: rolling constructively with neoliberalism? *Local Economy*, *30*(1), 98–118. https://doi.org/10.1177/0269094214563915
MoT. (2020). *Upper North Island Supply Chain Strategy*. Retrieved on 17 September 2020 from https://www.transport.govt.nz/multi-modal/keystrategiesandplans/upper-north-island-supply-chain-strategy/
Mouffe, C. (2005). *On the Political*. Routledge.
Murphy, L. (2008). Third-wave gentrification in New Zealand: the case of Auckland. *Urban Studies*, *45*(12), 2521–2540. https://doi.org/10.1177/0042098008097106
Ng, A. K. Y., & Ducruet, C. (2014). The changing tides of port geography (1950–2012). *Progress in Human Geography*, *38*(6), 785–823. https://doi.org/10.1177/0309132513516178
New Zealand Government. (2014). Port Companies Act 1988 No 91 (as at 01 December 2014), Public Act—New Zealand Legislation. http://legislation.govt.nz/act/public/1988/0091/latest/whole.html?search=sw_096be8ed80f9a7c8_evironm_25_se&p=1
NZ Herald. (2011). Forward thinking: metal heaven. *NZ Herald*, August 11, 2011.
NZ Herald. (2018). "We're right here!" Council worried about Government plans for port. *NZ Herald*, November 28, 2018.
NZ Herald. (2019a). Politics of the port. *NZ Herald*, December 5, 2019.
NZ Herald. (2019b). The three locations explained. *NZ Herald*, December 4, 2019.
NZ Herald. (2019c). Who will save Queens Wharf now? *NZ Herald*, March 1, 2019.
NZ Herald. (2020). Port holes. *NZ Herald*, July 10, 2020.
NZIER. (2017). *Future of New Zealand's vehicle supply chain: the role of the Ports of Auckland*. Retrieved on 10 July 2020 from https://voakl.files.wordpress.com/2017/12/port-of-auckland-relocation-nzier-report.pdf
O'Callaghan, C., & Linehan, D. (2007). Identity, politics and conflict in dockland development in Cork, Ireland: European Capital of Culture 2005. *Cities*, *24*(4), 311–323. https://doi.org/10.1016/j.cities.2007.01.006
Olivier, D., & Slack, B. (2006). Rethinking the port. *Environment and Planning A: Economy and Space*, *38*(8), 1409–1427. https://doi.org/10.1068/a37421
Ou, Z., & Ma, G. (2017). Marginalisation of the Dan fishing community and relocation of Sanya fishing port, Hainan Island, China. *Island Studies Journal*, *12*(2), 143–158. https://doi.org/10.24043/isj.32
Panuku Development. (2014). *Wynyard Precinct: Urban Design Framework*. Retrieved on 03 July 2020 from https://www.panuku.co.nz/downloads/assets/4983/1/wynyard%20quarter%20urban%20design%20framework.pdf
Peart, R. (2017). A 'sea change' in marine planning: the development of New Zealand's first marine spatial plan. *Policy Quarterly*, *13*(2), 3–9.

Peart, R. (2019). Sea Change Tai Timu Tai Pari: addressing catchment and marine issues in an integrated marine spatial planning process. *Aquatic Conservation: Marine and Freshwater Ecosystems*, 1–13. https://doi.org/10.1002/aqc.3156

POAL. (2018). *We Have a Plan: To Continue Delivering for Aucklanders*. Retrieved on 17 February 2022 from https://poal.maps.arcgis.com/apps/Cascade/index.html?appid=de22907ddb784d26bf60639763828d2d

POAL. (2019). *POAL—Ports of Auckland Releases Reviews of Upper North Island Supply Chain Study*. Retrieved on 10 July 2020 from http://www.poal.co.nz/media-publications/Pages/Ports-of-Auckland-releases-reviews-of-Upper-North-Island-Supply-Chain-study.aspx

POAL. (2020a). *Sparky! Auckland's New Electric Tug*. Retrieved on 11 July 2020 from http://www.poal.co.nz/media-publications/Pages/Sparky!-Auckland%E2%80%99s-new-electric-tug.aspx

POAL. (2020b). *World's Largest Soil-Based Vertical Garden Goes Up on Auckland Waterfront*. Retrieved on 11 July 2020 from http://www.poal.co.nz/media-publications/Pages/World%E2%80%99s-largest-soil-based-vertical-garden-goes-up-on-Auckland-waterfront.aspx

POAL. (2022). *Corporate Governance Statement*. Retrieved on 09 February 2022 from https://www.poal.co.nz/about-us/Pages/Corporate-Governance-Statement.aspx

Pyvis, J., & Tull, M. (2017). Institutions and port performance: a case study of the Port of Tauranga, New Zealand. *International Journal of Maritime History*, *29*(2), 276–306. https://doi.org/10.1177/0843871417692956

Robbins, P. (2012). *Political Ecology: A Critical Introduction* (2nd ed.). Wiley-Blackwell.

SKP. (2017). *Save Kennedy Point*. Retrieved on 7 July 2020 from www.skp.org.nz

Slack, B. (1980). *Technology and Seaports in the 1980s*. Blackwell Publishing Ltd.

Slack, B. (1989). The port service industry in an environment of change. *Geoforum*, *20*(4), 447–457. https://doi.org/10.1016/0016-7185(89)90028-6

Smellie, P. (2019). 'Surprising and 'concerning' conclusions in report on moving Auckland port. *NZ Herald*, November 21, 2019.

Sustainable Seas. (2020). *National Science Challenges: Sustainable Seas Ko ngā moana whakauka: Our Research*. Retrieved on 27 August 2020 from https://www.sustainableseaschallenge.co.nz/our-research/

Suykens, F. (1989). The city and its port—An economic appraisal. *Geoforum*, *20*(4), 437–445. https://doi.org/10.1016/0016-7185(89)90027-4

Tafon, R. V. (2018). Taking power to sea: Towards a post-structuralist discourse theoretical critique of marine spatial planning. *Environment and Planning C: Politics and Space*, *36*(2), 258–273. https://doi.org/10.1177/2399654417707527

UN-Habitat. (2018). *Blue Economy and Cities*. Retrieved on 3 September 2020 from https://unhabitat.org/sites/default/files/2020/04/un-habitat-background-paper-on-blue-economy-and-cities.pdf

Wakefield, S. (2007). Great expectations: waterfront redevelopment and the Hamilton Harbour Waterfront Trail. *Cities*, *24*(4), 298–310. https://doi.org/10.1016/j.cities.2006.11.001

Winder, G. M. (2006). Seafarer's gaze: Queen Street business and Auckland's archipelago, 1908. *New Zealand Geographer*, *62*(1), 50–64. https://doi.org/doi:10.1111/j.1745-7939.2006.00048.x

Winder, G. M., & Le Heron, R. (2017). Assembling a Blue Economy moment? Geographic engagement with globalizing biological-economic relations in multi-use marine environments. *Dialogues in Human Geography*, 7(1), 3–26. https://doi.org/10.1177/2043820617691643

13

BLUE ECONOMY POLICIES IN THE EUROPEAN UNION

The Case of French Maritime Clusters

Ebru Aricioğlu and Ezgi Biçer Uçar

13.1 Introduction

The Blue Economy, which is a component of sustainable development, has brought a new perspective to economic development. Although "Blue Economy" is an alternative development strategy introduced in the 2000s, unfortunately there is no unique agreed definition. Blue Economy offers many development opportunities to littoral countries, whether developed or developing. However, the models of production prevailing in today's economy have rapidly destroyed the environment including seas, oceans, lakes, etc. Hence, despite the benefits and new opportunities, the Blue Economy also includes crucial challenges for countries and regions in terms of sustainability. The way to manage these challenges varies from one country to another. Focusing on the background of the concept and denoting the contribution of Blue Economy to sustainable development, this chapter researches policies implemented in the European Union (EU) in order to promote Blue Economy. In this context, the aim of this chapter is to review the state of art policies about the Blue Economy in the EU and to identify the contributions of the Blue Economy industries to the EU economy via the case of France. Thereby, the study will contribute to the relevant literature by clarifying the concept of Blue Economy, which has no agreed definition yet, through a case study analysis and by concentrating mainly on the Blue Growth Clusters in the EU as a tool for the Marine Spatial Planning (MSP). In this frame, the Blue Growth Clusters in France will be highlighted. Within this purpose, first of all, the scope of Blue Economy and EU's path to the Blue Economy is presented as an output of the Integrated Maritime Policy (IMP) in the EU. As clusters fit well into the Blue Growth philosophy and its implementation (especially they are engaged as partners in MSP), the role played by the clusters in Blue Economy will be presented, with the special focus on the case of French Maritime Clusters.

DOI: 10.4324/9781003280248-17

Since the clusters in the maritime sector become significant to handle potential challenges in France.

13.2 The European Union's Path to the Blue Economy

Over the last half century, the definition of development and strategies to achieve it progressed and broadened from the maximisation of GDP in the 1950s, to employment creation and the satisfaction of basic needs in the 1970s, to structural adjustment and stabilisation in the 1980s and early 1990s, to poverty reduction, followed by sustainable and shared growth that dominated the scene until recently. The evolution in the conception of development culminated with the present broad-based concept of inclusive and sustainable growth. An important contribution of inclusive growth is that it recognises that human development is a highly multi-dimensional concept. Thus, development evolved from an essentially scalar concept to a multi-dimensional one entailing the simultaneous achievement of multiple objectives (Thorbecke, 2020: 101).

Development economics has followed a time path that moved it to become more experimental, more multidisciplinary, more rigorous, and more scientific. Some of the forces that influenced this evolutionary path are exogenous and some are endogenous. Researchers and the development community, in general, respond to socio-economic changes and to conceptual breakthroughs (Thorbecke, 2020: 102). Like the development economics, the perspective of institutions and governments and the policies they implement have also changed and transformed. The EU, UN, and the United States can be counted among these institutions and governments. It is worth noting that the 2008 financial crisis, which deeply affected the whole world, played a crucial role in this perspective change. With this change and transformation, the concepts of sustainable development, green economy, and Blue Economy began to take place in the agenda and policies of countries and institutions.

One of these aforementioned notions, the Blue Economy, is relatively recent, having become more widely used since the United Nations Conference on Sustainable Development in Rio de Janeiro in June 2012 (EU, 2020). Aiming at incorporating ocean values and services into economic modelling and decision-making processes, Blue Economy conceptualises oceans as "Development Spaces" and it constitutes a sustainable development framework for developing countries addressing equity in access to, development of, and the sharing of benefits from marine resources, offering scope for re-investment in human development and the alleviation of crippling national debt burdens. Hence, it breaks the mould of the business as usual "brown" development model where the oceans have been perceived as a means of free resource extraction and waste dumping, with costs externalised from economic calculations (UN, 2014).

There has been increasing interest in the Blue Economy around the world, yet there is no accepted definition of Blue Economy (Choi, 2017; Silver et al., 2015; Winder & Le Heron, 2017). Although it has received increased attention

TABLE 13.1 The course of Blue Economy policies of the European Union

2021	• A new approach for a sustainable blue economy in the EU
2014	• Innovation in the blue economy: realising the potentialof the seas and oceans for jobs and growth
2012	• Blue growth-oppurtunities for marine and maritime sustainable growth
2007	• An integrated maritime policy for the EU
2005	• Strategic Objectives 2005–2010 Europe 2010

Source: https://ec.europa.eu/

from the public and policymakers, interpretation of the term differs from one organisation to another as indicated by Emma McKinley in the introduction chapter of the book.

The EU defines the Blue Economy, which is one of the building blocks of its sustainable development strategy, as follows: the term Blue Economy simply refers to any economic activity relating to oceans and seas, and policy efforts are then targeted at "greening the Blue Economy." This definition covers a broad range of established and emerging sectors. For example, on energy, it includes both the traditional offshore extraction of the oil and gas sector and renewable energy sectors such as the established offshore wind industry and the emerging ocean energy industry (EU, 2020).

Table 13.1 shows the course of the EU's Blue Economy policies. In 2005, The European Commission, in its "Strategic Objectives 2005–2009—Europe 2010: A Partnership for European Renewal: Prosperity, Solidarity and Security" stated that in view of the environmental and economic value of the oceans and the seas, there is a particular need for an all-embracing maritime policy aimed at developing a thriving maritime economy and the full potential of sea-based activity in an environmentally sustainable manner. Such a policy should be supported by excellence in marine scientific research, technology, and innovation. As the relationship between sustainable development and the Blue Economy has not been explained in detail in this report, it is possible to interpret this document as a declaration of intent.

The EU's IMP, set up in 2007, is a framework to cover the entire maritime economy and to develop coordinated, coherent, and transparent decision-making in relation to all policies affecting the oceans, seas, islands, coastal and outermost regions, and maritime sectors. Aware of the need for a reinforced cooperation and effective coordination of all sea-related policies at the different decision-making levels while applying such an integrated, inter-sectoral approach; the governance framework and cross-sectoral tools necessary for an EU IMP have also been set in this communication. MSP has been introduced as a fundamental tool for the sustainable development of marine areas and coastal regions and for the restoration of Europe's seas to environmental health (COM, 2007). As a comprehensive approach to managing uses of the sea, MSP is defined as public process of analysing and allocating the spatial and temporal distribution of human activities in marine areas and is declared among the Ecosystem-Based Management Strategies in order to promote a Blue Economy and the sustainable development of maritime activities (www.mspglobal2030.org).

Within this wider scope, in 2012 the Commission launched its blue growth strategy, which forms the current framework for stimulating growth in the Blue Economy (EU, 2020). The EU evaluates the Blue Economy as an opportunity to get out of the crisis and to ensure the growth of the economy. Moreover, in the communication report, it is emphasised that sea and coasts are drivers of the economy, because their outward-looking geography, ports, and coastal communities have traditionally been centres for new ideas and innovation. In addition to this traditional propensity for innovation, three new factors have now come into play. First, there has been rapid technological progress in working offshore in ever deeper waters. Second, we are increasingly aware that land and freshwater are finite resources. Third, the need to reduce greenhouse gas emissions has not only driven the deployment of offshore renewable energy installations but has also provided a further impetus for energy saving and an additional reason to favour seaborne transport over land transport due to its lower emissions per tonne-kilometre.

This has opened an opportunity for blue growth—an initiative to harness the untapped potential of Europe's oceans, seas, and coasts for jobs and growth. Blue growth represents the maritime dimension of the Europe 2020 Strategy. It can contribute to the EU's international competitiveness, resource efficiency, job creation, and new sources of growth whilst safeguarding biodiversity and protecting the marine environment, thus preserving the services that healthy and resilient marine and coastal ecosystems provide (EU, 2020).

In attempts to encourage blue growth in the EU, new legislation was adopted in 2014, by the so-called Maritime Spatial Planning Directive. This directive is accommodating other EU directives and communications, such as the Marine Strategy Framework Directive and the Integrated Framework Policy (IMP) (Soma et al., 2018: 364). Additionally in 2014, the Blue Economy Innovation Plan was launched, specifying that the plan will be executed from three aspects: (I) develop sectors that have a high potential for sustainable jobs and growth; (II) essential components to provide knowledge, legal certainty, and security in the Blue Economy; and (III) sea basin strategies to ensure tailor-made measures and to foster cooperation between countries. Blue growth can also be considered as an extension of the land-based policy strategy referred to as green growth, which the EU has introduced in 2010.

In 2017, the EU issued the Report on the Blue Growth Strategy Toward More Sustainable Growth and Jobs in the Blue Economy. This report examines what has been learned and what has been achieved since 2012, what is ongoing, and what is still missing. Five aspects are described in the report:

1. pushing for growth in five focus areas, including blue energy, aquaculture, coastal and maritime tourism, blue biotechnology, seabed mineral resources,

2. The benefits of marine data, spatial planning, and maritime surveillance to facilitate growth in the Blue Economy,
3. promoting a partnership approach,
4. boosting investment, and
5. making blue growth strategy fit future challenge (Wenhai et al., 2019: 3) In the report, it is emphasized that "in our quest for natural resources and economic growth, we do not repeat the same mistakes on sea as we did on land. Blue growth is still in its early stages. Our responsibility today is to make sure that maritime economic development leads to a sustainable and competitive Blue Economy" (EU, 2017).

When the policies implemented by the EU are evaluated in general, Garland et al. (2019) emphasised that the EU firstly recognised the Blue Economy as a paradigm for regional economic growth in 2012. In defining the "boundaries" of the Blue Economy, the EU framed this clear regional economic paradigm within the boundaries of environmental sustainability. Garland et al. (2019) also emphasised that in the documents produced by the EU (and the United States), principles of sustainability, environmental justice, and equity do not actively inform or shape the selection or prioritisation of industrial or economic sectors considered part of Blue Economy. The EU (and the United States) perspective of the Blue Economy is therefore characterised by a lack of an integration of the sustainability and social justice principles. However, their narratives of ocean governance and maritime security still consist of peaceful and placeless representations, and this can misconstrue proper and effective sustainable ocean management (Garland et al., 2019).

By 2021, the EU changed its policy about the Blue Economy "making the transition from blue growth to a sustainable Blue Economy," and in this communication report, it is stressed that the EU needs to better connect the green and the blue policies, while extending its approach beyond EU borders and leading the way on international ocean governance and EU borders and leading the way on international ocean governance. Transforming the EU's Blue Economy for sustainable future requires a transformation of the value chains (COM, 2021). In this context, five hereinbelow initiatives have been put forward:

- Achieving the objectives of climate neutrality and zero pollution
- Circular economy and preventing waste
- Biodiversity and investing in nature
- Coastal resilience
- Responsible food systems

Moreover, in the same way as in the EU's other reports, tangible opportunities for new jobs and businesses creation potential of the Blue Economy are particularly emphasised (COM, 2021).

The Blue Economy Strategy highlights the fact that the Blue Economy needs to be sustainable and respect potential environmental concerns given the fragile nature of the marine environment. It focuses on five sectors selected following a study analysing the potential for innovation and job creation in the Blue Economy and where additional effort at EU level could stimulate long-term growth and jobs in the Blue Economy: "blue" energy, aquaculture, coastal and maritime tourism, "blue" biotechnology, and marine mineral resources. In the EU Blue Economy Report 2019 (EU, 2019), Blue Economy established sectors are defined as, marine living resources, marine extraction of non-living resources, maritime transport, port activities, shipbuilding and repair, and coastal tourism. Blue energy (like offshore wind energy), blue bio-economy and biotechnology, and marine minerals are cited as the emerging and innovative sectors of Blue Economy. Some research categorises the established sectors as "Blue Economy" and the emerging sectors as the "Blue Growth." In both cases, the Blue Economy is linked to many other economic activities and its impact goes beyond these sectors (EC, 2020). It is possible to point out that the "Blue Growth" was used to discover the untapped potential of Europe's oceans, seas, and costs for jobs and growth (De Vet et al., 2016).

According to the most recent figures, the established sectors of the EU Blue Economy directly employed close to 5 million people and generated around €750 billion in turnover and €218 billion in Gross Value Added (GVA) in 2018. The relative size of the EU Blue Economy in terms of GVA with respect to the overall economy has remained stable at around 1.5% since 2012, while it has increased in terms of employment from 1.8% in 2015 to more than 2.2% in 2018. In terms of GVA and employment, coastal tourism has the highest share (Blue Economy Report, 2020). As seen in Table 13.2, in all socioeconomic indicators, marine renewable energy has the highest growth rate so this fact shows the importance of new sectors as well as established sectors. Facing many downside risks (financial crisis and economic recession) and uncertainties (coronavirus pandemic) in the world economy, the Blue Economy occurs as a promising response to these negative factors and proposes a solid base for further development.

13.3 Clusters in Blue Economy: The case of French Clusters

As the world moves toward the implementation of the 2030 Agenda for Sustainable Development, more than ever MSP is considered as an effective framework to guide the sustainable development of the ocean and coasts and achieve global ocean governance goals.

MSP is also a common denominator for maritime clusters, as spatial planning along with functioning ecosystems are a prerequisite for the long-term sustainable development of the maritime clusters. As activities at sea or near shore may conflict mutually or with environmental objectives, spatial planning is needed to

TABLE 13.2 Growth rate of total established sectors between 2009 and 2018

Growth rate %Δ 2009–18	*Total of established sectors*	*Marine living resources*	*Marine non-living resources*	*Marine renewable energy*	*Port activities*	*Shipbuilding-repair*	*Maritime transport*	*Coastal tourism*
Employment	0.9%	−3.1%	−28.4%	693.3%	20.3%	−9.7%	7.2%	−0.3%
Value added at factor cost	15.0%	23.7%	−28.8%	1,276.3%	24.0%	32.3%	19.2%	19.7%
Net investment	−7.7%	−6.8%	−11.9%	1,268.1%	−41.5%	−19.5%	–	–
Turnover	11.7%	26.4%	−59.0%	1,185.0%	38.7%	13.4%	30.6%	17.6%
Personnel costs	13.4%	16.2%	−22.2%	819.9%	30.5%	14.6%	18.5%	9.5%

Source: Eurostat

reduce such conflicts and the potential impact. Maritime clusters are well placed to engage as partners in MSP and to provide a wealth of information on future space needs. Such exchange is also an opportunity to strengthen the consciousness of economic actors about the need for an ecosystem-based approach (DG MARE,[1] 2014: 68–69).

As noted above, clusters fit well into the Blue Growth philosophy and its implementation (DG MARE, 2014: 11). The cluster phenomenon has taken on an important dimension in regional development strategies since the 1990s. Defining clusters as geographic concentrations of interconnected companies and institutions in a particular field, Porter stresses that clusters encompass an array of linked industries and other entities important to competition (Porter, 1998: 78). Porter also adds that: "Each regional economy normally has a relatively small number of traded clusters *in which the region is truly competitive with other regions and countries. … There is no national economy, then, but a series of regional economies that trade with each other and the rest of the world, each with its own pattern of cluster specialization. Such regional specialization drives productivity and productivity growth in the national economy. There is growing statistical evidence that regions with stronger clusters achieve better economic performance and faster innovation*" (Porter, 2007: 2). As soon as we try to understand the different forms that a cluster can take depending on their composition, function, and governance (Shi et al., 2020: 239), we notice that the trajectories of development and integration into regions vary from cluster to cluster. When the EU's perspective on clusters is analysed in more detail, the Union considers clusters as a tool to gain competitive advantage. Moreover, in the Communication of the Commission, lack of access to finance and a shortage of suitably skilled workers have been identified as blocking growth in nearly all economic sectors. It is indicated that in the Blue Economy, Member States are tackling this by developing "**maritime clusters**." These are groupings of larger industries, smaller suppliers, and educational establishments that reinforce each other through their proximity. The ship-repair business in Brest which plays host to France's largest maritime cluster, the "Pôle de Competitivité Mer," is quoted as an example (COM, 2012: 4). The importance attached by the EU to the role of the maritime clusters to enhance the strength and development in European maritime sectors took place in the previous publications of the Union as well (DG MARE, 2008). While using the "maritime cluster" term, EU documents define what they mean by a maritime cluster. Nevertheless, in the literature, there is also a debate about what a maritime cluster is. As the maritime clusters rapidly became the focus of economic competitiveness policies, this question became parallelly important. The definition of the EU shows us that the EU considers maritime cluster as a community-based network rather than as an industrial complex or an agglomeration of interlinked industries (Doloreux, 2017: 215–216). We face a similar challenge to make a definition of a maritime cluster, like in defining the "Blue Economy."

In 2007, Shearmur et al. mention about the relative scarcity of the analysis carried out to better understand the cluster phenomenon in the maritime sector, particularly regarding the main mechanisms that explain their development and functioning, as well as the development trajectories that characterise them and their integration into regional economies (Shearmur et al., 2007: 366–367). Even though maritime clusters draw more attention in the recent years, the absence of consensus on "what is a maritime cluster" has increased the difficulty in further expansion of maritime clusters–related studies; and the limited number of maritime clusters and relevant customised data has hindered the verification, generalisation, and applicability of the conclusions drawn from the conventional research (Shi et al., 2020: 237–254).

Aware of the fact that a clear definition of the boundary of maritime clusters is critical to maritime cluster development in actual practice, particularly with measuring the cluster effect, configuring the cluster governance, and determining the policy framework (Shi et al., 2020), we would define the maritime clusters as geographic concentrations of firms in maritime sectors, of research and education organisation active in related fields and public support mechanisms operated by the government and regional stakeholders (Doloreux, 2017). Focusing on the maritime clusters in France, it should be reminded that France has the world's second largest maritime territory in terms of surface area. From fishing to ocean biotechnologies, half a million people are employed in the maritime economy, equivalent to 2% of overall national employment. On average, establishments operating in the maritime economy are smaller and younger than establishments in other sectors. Maritime employment is naturally concentrated in coastal zones, with half of these jobs in the Provence-Alpes-Côte d'Azur and Brittany regions (INSEE, 2015). These two regions host two competitive maritime clusters as well. European Cluster Collaboration Platform (https://clustercollaboration.eu/) shows us that France hosts 10 clusters dedicated to blue growth industries.[2] In this classification, the blue growth is a cross-sectoral industry. Looking at the performance of the EU countries based on the number of blue growth industry clusters that they host, with 14 clusters, Spain is the leading country, followed by France (10 clusters) and Italy (8 clusters) (Figure 13.1).

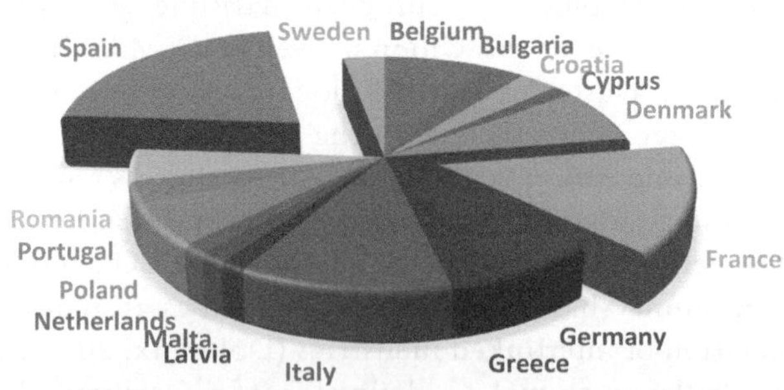

FIGURE 13.1 Blue growth industry clusters in EU 27

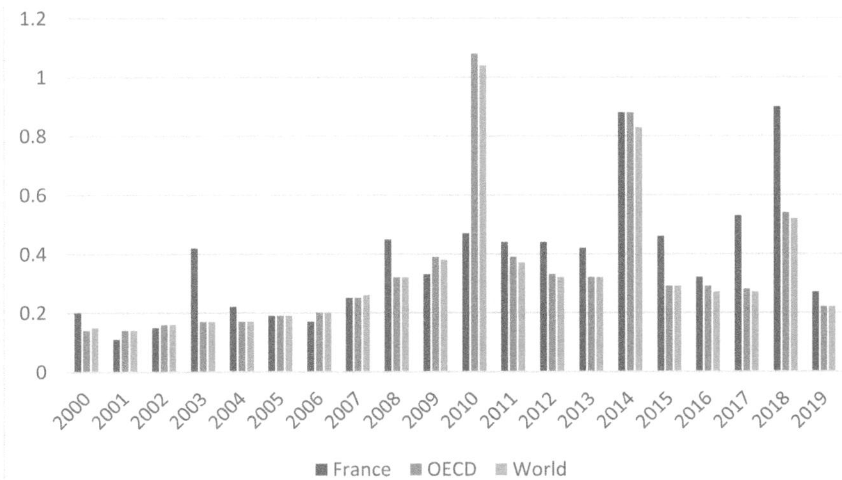

FIGURE 13.2 Ocean-related high-value ENVTECH inventions %

Figure 13.2 shows ocean-related high-value ENVTECH inventions as a share of total high-value inventions. When France is compared with OECD and World average, the ratio of patents of relevance to ocean sustainability, this rate seems to be higher in France. The role of clusters in this case is undeniable. Despite the performance of France in terms of maritime clusters, it is interesting to notice that there are no studies dedicated to France among the studies on maritime clusters mentioned in the appendix of Doloeux's article (Doloreux, 2017: 2018).

The French Maritime Cluster (Cluster Maritime Français [CMF]), founded in 2006 with a legal status of an association, aims at promoting and defending the maritime and related activities of France. It is also keeping with, in a quasi-ideal way, in terms of schedule as much as philosophy, the spirit and dynamics which the Green Paper (on and for an integrated maritime European policy) intends to promote. Vallat and Perennez (2006: 81) emphasised that according to 2006's data, the cluster of maritime industries employs about 1.5% of the French labour force while production, excluding the intermediary purchases within the cluster, amounts to between 2% and 2.5% of the gross domestic product. Moreover, even in restricting us to the maritime cluster in a narrow sense, it is important to note that about half of its production is exported. In money terms, it is at the same level as the exports of the French car industry or the aeronautical industry and at three times the level of exports of the famous French wine industry (Vallat & Perennez, 2006). Today, CMF is made up of more than 430 entities: companies of all sizes, competitiveness clusters, federations and associations, laboratories and research centres, schools and training organisations, local communities, and economic players, as well as the French Navy (EU, 2020: 17).

While the CMF is a player at the national vise, one of the competitive maritime clusters is Pôle de Compétitivité Mer (Pôle Mer Bretagne Atlantique) which covers all maritime sectors and blue growth while environmental industries are considered emerging industries of the cluster. The primary criterion for the emergence of a cluster is its ability to bring together large and small companies, research centres and laboratories, and training and education establishments within a specific region and around a common theme. Pôle Mer Bretagne Atlantique is a classic example of such a cluster in practice (www.pole-mer-bretagne-atlantique.com). Smart specialisation[3] priority areas are defined as aquaculture and blue renewable energy. Gathering 390 members, Pôle Mer hosts 220 SMEs and 30 of them are start-ups. While the number of larger companies is 45, there are also 52 research organisations, universities, and technology centres. So, it is possible to consider this cluster as a good example of governance. There is also another ambitious cluster in the same domain which is Pôle Mer PACA (Provence-Alpes-Côte d'Azur). These two French marine competitiveness clusters are coordinated across two major coastal regions: Brittany and PACA. These two regions have significant potential for industrial growth in the marine sector (700 companies), and 75% of French research in the field is concentrated here.

BRIT'INOV hosting 23 SMEs, BPN Brittany Naval Pole hosting 155 SMEs, CBB Capbiotek hosting 176 SMEs, and ID2Santé hosting 85 SMEs are the other clusters in the region who defined the blue growth industries as the emerging industries (www.clustercollaboration.eu). The agglomeration of the maritime clusters in Brittany is related with the 2,730 kilometres of coastline of the region which represents one-third of the total coastline of mainland France. Therefore, shipbuilding takes place among the key sectors with food, automotive, and Information and Communications Technology (ICT). Maritime economy and notably marine renewable energies, sustainable use of biomass and biotechnologies, recovery and use of marine resources, fishing and aquaculture, shipbuilding, and maritime security and safety are among the seven smart specialisation sectors of the region (www.ec.europa.eu/growth/tools-databases/regional-innovation-monitor/base-profile/bretagne). Besides the competitiveness and regional clusters, there are also initiatives, like Campus Mondial de la Mer, bringing together France's first community devoted to understanding and developing marine resources.

The aim of the community is to consolidate Brittany's position as one of the key places worldwide for the study and development of the oceans and seas, alongside major international centres of maritime excellence (www.campusmer.fr/files/2423/Dossier_Presse_CMM_uk.pdf).

13.4 Conclusion

The bottlenecks, recessions, and crises experienced by the world economy have shown the need for a new development paradigm. Since Blue Economy is a concept that goes beyond country borders and affects many variables at the same time and affects our future resources and growth, it is a phenomenon that needs to be governed. In fact, Blue Economy has two contradicting ingredients: protection of resources and economic growth. However, if Blue Economy is governed from a win-win perspective, it may be possible to increase national income, to create new jobs, and to create new investment areas.

Many countries consider Blue Economy to be the new development approach in this period. One of them is the EU. While every member of the EU doesn't attach the same level of importance to Blue Economy, the future French champions of the sea are considered by CMF as a real challenge for France, like Google or Apple today in the world of the Internet. Hosting the second largest Exclusive Economic Zone of the world, it is supposed that Blue Economy would sow the seeds for new jobs and for the rare development prospects of the overseas territories of France (www.cluster-maritime.fr). Even if the clear definition of the term "maritime cluster" doesn't exist such as the "Blue Economy," maritime clusters are widely used by governments as a strategic tool to promote maritime industries (Doloreux, 2017: 215), like in the case of France. However, the success of France in creating the future champions of the sea would depend on how the country would foster the private-public partnership (PPP) in the sector. For sure clusters present an ideal framework for PPP, however in the case of the Brest, as an important centre of maritime activity, the focus is on defence; therefore, the exchanges between public and private activities are still limited (Shearmur et al., 2007: 382). However, the potential of the PPP that could be flourished in the region can only be discovered by deeply analysing the actors of the various clusters via field research in the region, and this could be focus of further research.

Notes

1 Directorate-General for Maritime Affairs and Fisheries.
2 These clusters are as follows: ASTech Paris Region (Île-de-France), BPN Brittany Naval Pole (Brittany), BRIT'INOV (Brittany), Pôle Mer Bretagne Atlantique (Brittany), Cluster GAT CARAIBES Logistique et Transports (Martinique), EMC2 (Loire Region), Pôle Mer Méditerranée—Business & Innovation Sea Cluster (Provence-Alpes-Côte d'Azur), Riviera Yachting NETWORK (Provence-Alpes-Côte d'Azur), Éa éco-entreprises (Provence-Alpes-Côte d'Azur), and TEMERGIE (Réunion).
3 The notion especially describes the capacity of an economic system to generate new specialisms through use of "existing resources" (Foray, 2015: 1).

References

Choi, Y.R. (2017). "The Blue Economy as governmentality and the making of new spatial rationalities". *Dialogues in Human Geography*, 7 (1) 37–41.

COM. (2007). 575 Final, "An Integrated Maritime Policy for the European Union". Communication from the Commission to the European Parliament, the Council, the European Economic and Social Committee and the Committee of Regions.

COM. (2012). 494 Final, "Blue Growth". Communication from the Commission to the European Parliament, the Council, the European Economic and Social Committee and the Committee of the Regions.

COM. (2021). 240 Final, "On a new approach for a sustainable Blue Economy in the EU Transforming the EU's Blue Economy for a Sustainable Future". Communication from the Commission to the European Parliament, the Council, the European Economic and Social Committee and the Committee of the Regions.

De Vet, J.-M., Edwards, J., Bocci, M. (2016). Blue growth and smart specialisation: How to catch maritime growth through 'Value Nets', S3 Policy Brief Series No. 17/2016.

Directorate-General for Maritime Affairs and Fisheries (DG MARE) (2008). The Role of Maritime Clusters to Enhance the Strength and Development of European Maritime Sectors, Report on Results. Study done on behalf of the European Commission (DG MARE), Directorate-General for Maritime Affairs and Fisheries.

Directorate-General for Maritime Affairs and Fisheries (DG MARE) (2014). "Support activities for the development of maritime clusters in the Mediterranean and Black Sea". Final Report under FWC MARE/2012/06.

Doloreux, D. (2017). "What is a maritime cluster?". *Marine Policy*, 83, 215–220.

European Commission (EC) (2020). The EU Blue Economy Report, 2020. Brussels

EU. (2017). 128 Final, "Report on the Blue Growth Strategy towards more sustainable growth and jobs in the Blue Economy". Commission Staff Working Document, SWD (2017).

EU. (2019). "The EU Blue Economy report".

EU. (2020). "The EU Blue Economy report and report's annexes".

Foray, D. (2015). *Smart Specialisation Opportunities and Challenges for Regional Innovation Policy*, Routledge.

Garland, M., Axon, S., Graziano, M., Morrisey, J., Heidkamp, C.P. (2019) "The Blue Economy: Identifying geographic concepts and sensitivities". *Geography Compass*, 13(7), e12445.

Porter, M. (1998). "Clusters and the New Economics of Competition". *Harvard Business Review*, 76(6), 77–90. Reprint 98609.

Porter, M. (2007). "Clusters and economic policy: Aligning public policy with the new economics of competition". *Harvard Business School ISC White Paper*, November.

Shearmur, R., Chenard, P., Doloreux, D. (2007). "La création et le développement de clusters maritimes au Canada et en Europe". *Revue d'Économie Régionale & Urbaine*, 3(3). https://doi.org/10.3917/reru.073.0365

Shi, X., Jiang, H., Li, H., Xu, D. (2020). "Maritime cluster research: Evolutionary classification and future development". *Transportation Research Part A* 133, 237–254.

Silver, J.J., Gray, N.J., Campbell, L.M., Fairbanks, L.W., Gruby, R.L. (2015). "Blue Economy and competing discourses in international oceans governance". *Journal of Environment and Development*, 24(2). https://doi.org/10.1177/1070496515580797.

Soma, K., van den Burg, S.W.K., Hoefnagel, E.W.J., Stuiver, M., Martijn van der Heidea, C. (2018). "Social innovation—A future pathway for Blue growth?". *Marine Policy* 87, 363–370.

The French National Institute of Statistics and Economic Studies (INSEE) 2015. The maritime economy: diverse and geographically-specific activities. Paris.

Thorbecke, E. (2020). "The history and evolution of the development doctrine, 1950–2017". In: Machiko N., Ocampo, J.A., Macmillan, P. (Eds.), *The Palgrave Handbook of Development Economics Critical Reflections on Globalisation and Development*. Cham: Palgrave Macmillan.

United Nations. (2014). "The Blue Economy concept paper". *Sustainable Development Goals Knowledge Platform*.

Vallat, F., Perennez, P. (2006). "The French Maritime Cluster". In: Wijnolst, N. (Ed.), *Dynamic European Maritime Clusters*. Amsterdam: IOS Press B.

Wenhai, L., Cusack, C., Baker, M., Tao, W., Mingbao, C., Paige, K., Xiaofan, Z., Levin, L., Escobar, E., Amon, D., Yue, Y., Reitz, A., Neves, AAS., O'Rourke, E., Mannarini, G., Pearlman, J., Tinker, J., Horsburgh, KJ., Lehodey, P., Pouliquen, S., Dale, T., Peng, Z., Yufeng, Y. (2019). "Successful Blue Economy examples with an emphasis on international perspectives". *Frontiers in Marine Science*. https://doi.org/10.3389/fmars.2019.00261

Winder, G.M., Le Heron, R. (2017). "Assembling a Blue Economy moment? Geographic engagement with globalizing biological-economic relations in multi-use marine environments". *Dialogues in Human Geography*, 7 (1), 3–26.

14

THE EVOLUTION OF BLUE CARBON

Exploring the Burgeoning Role of Macroalgae in Carbon and Nitrogen Sequestration

Louie Krak and C. Patrick Heidkamp

14.1 Introduction

From 2005 to 2015, global seaweed production more than doubled, increasing from 14.7 million tons to 30.4 million tons in the span of a decade (Ferdouse et al., 2018). Seaweed is exceptionally versatile, with products ranging from food and cosmetics to fertiliser and animal feed, at a total annual value of production of 6 billion USD (McHugh, 2003). Beyond its substantial commercial potential, seaweeds and kelps have also garnered attention as possible sources of Blue Carbon (BC). BC, or carbon that is captured and sequestered by coastal and marine ecosystems (Lovelock and Duarte, 2019), is a term that was first coined in 2009 and has since acquired multiple meanings that include all of the organic matter that marine organisms capture as well as the management of marine ecosystems to reduce greenhouse gas emissions (Lovelock and Duarte, 2019).

Traditionally, the BC discourse has focused on ecosystems such as mangroves, seagrasses, and tidal marshes, which collectively sequester carbon not only within their living aboveground and belowground biomass but also in their underlying sediments (Mcleod et al., 2011). These ecosystems have gained growing international interest for their prodigious carbon sequestration capacity. Although these three BC habitats cover just a small fraction of the global seafloor, they capture and store an estimated 11.5 billion tons of carbon (Ahmed et al., 2016). These BC ecosystems are important not only for their ability to mitigate greenhouse gas emissions by capturing and sequestering carbon dioxide (CO_2) from the air but also because their conservation, restoration, and protection prevent CO_2 emissions from being released as a result of their destruction (Lovelock and Duarte, 2019).

Although mangroves, seagrasses, and tidal marshes remain the traditional sources of BC, other potential BC sources such as macroalgae have recently been considered in the BC discourse. An inventory of BC literature was first

DOI: 10.4324/9781003280248-18

conducted by Thomas in 2014. This meta-analysis spanned from the inception of BC in 2009, tracking its trajectory through July of 2013 (Thomas, 2014). Thomas identified both peer-reviewed and "grey literature" documents that focused on BC as it pertains to finance and market mechanisms (Thomas, 2014). Utilising both quantitative and qualitative methods, Thomas's study revealed that the BC discourse contains predominantly technical and policy commentary that lacks practical, social, and private sector perspectives (Thomas, 2014). This investigation seeks to provide a quantitative and qualitative examination of the BC literature that has been published since Thomas's meta-analysis in order to (1) argue for the inclusion of macroalgae as a non-traditional BC ecosystem, (2) discuss the comparatively underdeveloped literature on BC carbon offsetting, (3) emphasise the importance of a wholistic approach to payment for ecosystem services (PES), and (4) introduce the concept of "Blue Nitrogen."

While the concept and definitions of the Blue Economy are vague and fluid, there is agreement among many (see for example McKinley, chapter 2; Axon, chapter 4; Lewis and LeHeron, chapter 8; this volume) he authors of the contributions to this edited volume) that it is a new development paradigm which is focused on the sustainable and equitable use of ocean resources. In order to do so, the Blue Economy as a paradigm will need to consider a shift towards a low-carbon economy, a shift towards a shared prosperity among the worlds' people, all while maintaining the functioning of the bio-physical system. Given that BC, as outlined by Steven et al. (2019), can be considered (1) a market-based mechanism to move to low-carbon economies, (2) a creator of economic value with a significant potential, particularly in developing countries, and (3) a provider of important ecosystem services in the coastal and marine environment. Thus, there is a clear link between the concept of BC and the emerging Blue Economy. "Blue Carbon meets all of three of the essential elements of sustainability—economic, environmental and social—and should thus be framed as an essential component of the Blue Economy" and is an emerging Blue Economy sector in its own right (Steven et al., 2019).

14.2 Methods

The two-part analysis began by examining the relevant BC literature that has been published since the conclusion of Thomas's (2014) literature search in July 2013. In the systematic literature review, 38 papers were selected that engaged with the recent BC discourse. Articles that focus on traditional BC ecosystems such as mangroves, seagrasses, and tidal marshes were included in addition to others that outlined non-traditional sources like macroalgae, seaweeds, and kelps. This meta-analysis identified research papers related to BC that were English language, predominantly peer-reviewed, and scientific in nature. These articles were found using online databases such as EBSCO, Google Scholar, and Mendeley, and they collectively provide interdisciplinary perspectives spanning biology, economics, geography, and policy.

The literature search was conducted over an extended period from March 2020 through June 2020 using "Blue carbon" as the primary keyword in combination with other search terms such as "Seaweed," "Macroalgae," "Kelp," "Sequestration," and "Carbon offsetting." Once the selection of papers was finalised, the 38 articles were coded using the following metrics: authorship, year of publication, article type, journal source, study location, species studied, and the main topics addressed in each paper (scientific or technical, social, economic, regulatory or policy, and management). The results of the coding are displayed in Table 14.1.

Second, the literature was analysed using the NVivo 12 Plus software package that was released in March 2020. NVivo is a data organisation and management tool that provides qualitative analysis via thematic coding, and it is utilised for research in academia, government, the non-profit sector, health, and commercial enterprises (NVivo, 2020). The software performed automatic thematic coding that detected and tabulated the themes present in the imported documents. The "automated insights" feature assigned broad categorisations to the data and provided an objective thumbprint of the patterns or trends in the literature (NVivo, 2020).

When applied to the 38 articles selected for this investigation, the automated insights auto-code feature generated 5 overarching themes and 1,125 subthemes using the Auto Code Wizard function. Within each overarching theme, the subthemes were manually aggregated into categories with 25 or more references per subtheme. After the manual aggregation, the overarching subthemes with 25 or greater references were each individually subjected to a Text Search Query (TSQ). Each word or phrase was entered into the TSQ, and the software tabulated the number of references in each article and then summed the total number of references across all 38 articles. The TSQ took "stemmed words" into account such that searching for the term "sequester" would also include variations such as "sequestered" and "sequestering" in the final reference count.

Separate TSQs were performed for singular and plural forms of search words and phrases to account for variations such as "carbon cycle" and "carbon cycles." TSQs were also run for abbreviations such as "CO_2" in place of "carbon dioxide." The singular, plural, and abbreviated references were all summed to generate the total number of references for each overarching theme and subtheme, and the results were displayed in the treemap in Figure 14.1. An additional overarching theme was added to broaden and contextualise the scope of the analysis. The "Miscellaneous" overarching theme included subthemes that were selected both from the results of a Word Frequency Query (WFQ) and additional terms that were selected. The results of the WFQ can be found in Figure 14.2.

The WFQ searched the 38-article set and identified the 20 most frequently used words in the entire body of literature. References to "Blue," "Carbon," and any other stop words that were deemed as clutter were removed. Stemmed words were included to account for different variations of the same word. For example, the words "Climate," "Climates," and "Climatic" were all accounted for in the total word count for "Climate." Even though the term "Nitrogen" was not in

TABLE 14.1 Results of the systematic quantitative literature review

							Scientific			Regulatory	
Author	*Year*	*Title*	*Type*	*Journal*	*Study location*	*Species*	*Technical*	*Social*	*Economic*	*Policy*	*Management*
Ahmed, N., Bunting, M.G., Flaherty, M.S., Diana, J.S.	2016	Can greening of aquaculture sequester blue carbon?	Peer-reviewed article	*Ambio*	Global	Mangroves, seagrasses, salt marshes, seaweeds					X
Alongi, D.M., Murdiyarso, D., Fourqurean, J.W., Kauffman, J.B., Hutahaean, A., Crooks, S., Lovelock, C.E., Howard, J., Herr, D., Fortes, M., Pidgeon, E., Wagey, T.	2015	Indonesia's blue carbon: a globally significant and vulnerable sink for seagrass and mangrove carbon	Peer-reviewed article	*Wetlands Ecology and Management*	Indonesia	Mangroves, seagrasses	X				
Campbell, I., Macleod, A., Sahlmann, C., Neves, L., Funderud, J., Overland, M., Hughes, A.D., Stanley, M.	2019	The Environmental Risks Associated with the Development of Seaweed Farming in Europe—Prioritizing Key Knowledge Gaps	Peer-reviewed article	*Frontiers in Marine Science*	Europe	Seaweeds, kelp	X			X	X
Campbell, J.E., Lacey, E.A., Decker, R.A., Crooks, S., Fourqurean, J.W.	2015	Carbon Storage in Seagrass Beds of Abu Dhabi, United Arab Emirate	Peer-reviewed article	*Estuaries and Coasts*	United Arab Emirates	Seagrasses	X				

(Continued)

TABLE 14.1 Results of the systematic quantitative literature review *(Continued)*

							Scientific			*Regulatory*	
Author	*Year*	*Title*	*Type*	*Journal*	*Study location*	*Species*	*Technical*	*Social*	*Economic*	*Policy*	*Management*
Chung, I.K., Sondak, C.F.A., Beardall, J.	2017	The future of seaweed aquaculture in a rapidly changing world	Peer-reviewed article	*European Journal of Phycology*	Global	Seaweeds	X			X	
Duarte, C.M.	2017	Reviews and syntheses: Hidden forests, the role of vegetated coastal habitats in the ocean carbon budget	Peer-reviewed article	*Biogeosciences*	Global	Mangroves, seagrasses, salt marshes, macroalgae	X				
Duarte, C.M., Wu, J., Xiao, X., Bruhn, A., Krause-Jensen, D.	2017	Can Seaweed Farming Play a Role in Climate Change Mitigation and Adaptation?	Peer-reviewed article	*Frontiers in Marine Science*	Global	Seaweeds	X				X
Elania, E., Radiarta, I.N.	2014	Management of Sustainable Seaweed (*Kappaphycus alvareziiI)* Aquaculture in the Context of Climate Change Mitigation	Peer-reviewed article	*Indonesian Aquaculture Journal*	Indonesia	Seaweeds	X				

(Continued)

TABLE 14.1 Results of the systematic quantitative literature review *(Continued)*

							Scientific			*Regulatory*	
Author	*Year*	*Title*	*Type*	*Journal*	*Study location*	*Species*	*Technical*	*Social*	*Economic*	*Policy*	*Management*
Froehlich, H.E., Afflerbach, J.C., Frazier, M., Halpern, B.S.	2019	Blue Growth Potential to Mitigate Climate Change through Seaweed Offsetting	Peer-reviewed article	*Current Biology*	Global	Seaweeds	X			X	
Greiner, J.T., McGlathery, K.J., Gunnell, J., McKee, B.A.	2013	Seagrass Restoration Enhances "Blue Carbon" Sequestration in Coastal Waters	Peer-reviewed article	*PLoS One*	The United States	Seagrass (*Zostera marina*)	X				
Herr, D., von Unger, M., Laffoley, D., McGivern, A.	2017	Pathways for implementation of blue carbon initiatives	Peer-reviewed article	*Aquatic Conservation: Marine and Freshwater Ecosystems*	Global	Mangroves, salt marshes, seagrasses				X	X
Hill, R., Bellgrove, A., Macreadie, P.I., Petrou, K., Beardall, J., Steven, A., Ralph, P.J.	2015	Can macroalgae contribute to blue carbon? An Australian perspective	Peer-reviewed article	*Limnology and Oceanography*	Australia	Mangroves, seagrasses, salt marshes, macroalgae	X				

(Continued)

TABLE 14.1 Results of the systematic quantitative literature review *(Continued)*

Author	Year	Title	Type	Journal	Study location	Species	Scientific			Regulatory	
							Technical	Social	Economic	Policy	Management
Howard, J., Sutton-Grier, A., Herr, D., Kleypas, J., Landis, E., Mcleod, E., Pidgeon, E., Simpson, S.	2017	Clarifying the role of coastal and marine systems in climate mitigation	Peer-reviewed article	*Frontiers in Ecology and the Environment*	Global	Mangroves, seagrasses, tidal marshes	X			X	X
Iñiquez, C., Carmona, R., Lorenzo, M.R., Niell, F.X., Wienecke, C., Gordillo, F.J.L.	2016	Increased temperature, rather than elevated CO2, modulates the carbon assimilation of the Arctic kelps *Saccharina latissima* and *Laminaria solidungula*	Peer-reviewed article	*Marine Biology*	Norway	Kelp (*Saccharina latissima*) (*Laminaria solidungula*)	X				
Krause-Jensen, D., Duarte, C.M.	2016	Substantial role of macroalgae in marine carbon sequestration	Peer-reviewed article	*Nature Geoscience*	Global	Macroalgae	X				
Lavery, P.S., Mateo, M., Serrano, O., Rozaimi, M.	2013	Variability in the Carbon Storage of Seagrass Habitats and Its Implications for Global Estimates of Blue Carbon Ecosystem Service	Peer-reviewed article	*PLoS One*	Australia	Seagrasses	X		X		

(Continued)

TABLE 14.1 Results of the systematic quantitative literature review *(Continued)*

							Scientific			*Regulatory*	
Author	*Year*	*Title*	*Type*	*Journal*	*Study location*	*Species*	*Technical*	*Social*	*Economic*	*Policy*	*Management*
Locatelli, T., Binet, T.., Kairo, J.G., King, L., Madden, S., Patenaude, G., Upton, C., Huxham, M.	2014	Turning the Tide: How Blue Carbon and Payments for Ecosystem Services (PES) Might Help Save Mangrove Forests	Peer-reviewed article	*Ambio*	Global	Mangroves	X	X	X	X	X
Lovelock, C.E., Duarte, C.M.	2019	Dimensions of Blue Carbon and emerging perspectives	Peer-reviewed article	*Biology Letters*	Global	Mangroves, seagrasses, tidal marshes, macroalgae	X			X	X
Macreadie, P.I., Nielsen, D.A., Kelleway, J.J., Atwood, T.B., Seymour, J.R., Petrou, K., Connolly, R.M., Thomson, A.C.G., Trevathan-Tackett, St.M., Ralph, P.J.	2017	Can we manage coastal ecosystems to sequester more blue carbon?	Peer-reviewed article	*Frontiers in Ecology and the Environment*	United States, Australia	Mangroves, seagrasses, tidal marshes	X				X

(Continued)

TABLE 14.1 Results of the systematic quantitative literature review *(Continued)*

							Scientific			*Regulatory*	
Author	*Year*	*Title*	*Type*	*Journal*	*Study location*	*Species*	*Technical*	*Social*	*Economic*	*Policy*	*Management*
Macredie, P.I., Hughe, R.A., Kimbro, D.L.	2013	Loss of "Blue Carbon" from Coastal Salt Marshes Following Habitat Disturbance	Peer-reviewed article	*PLoS One*	The United States	Seagrass (*Thallasia testudinum*), salt marsh (*Spartina alterniflora*)	X				
Macreadie, P.I., York, P.H., Sherman, C.D.H., Keough, M.J., Ross, D.J., Ricart, A.M., Smith, T.M.	2014	No detectable impact of small-scale disturbances on "blue carbon" within seagrass beds	Peer-reviewed article	*Marine Biology*	Australia	Seagrass (*Zostera nigracaulis*)	X				
Macreadie, P.I., Baird, M.E., Trevathan-Tackett, S.M., Larkum, A.W.D., Ralph, P.J.	2014	Quantifying and modelling the carbon sequestration capacity of seagrass meadows—a critical assessment	Peer-reviewed article	*Marine Pollution Bulletin*	Global	Mangroves, seagrasses, salt marshes	X		X	X	X
Oreska, M.P.J., Wilkinson, G.M., McGlathery, K.J., Bost, M., McKee, B.A.	2018	Non-seagrass carbon contributions to seagrass sediment blue carbon	Peer-reviewed article	*Limnology and Oceanography*	The United States	Seagrass (*Zostera marina*)	X				

(Continued)

TABLE 14.1 Results of the systematic quantitative literature review *(Continued)*

Author	Year	Title	Type	Journal	Study location	Species	Scientific			Regulatory	
							Technical	*Social*	*Economic*	*Policy*	*Management*
Oreska, P.J., McGlathery, K.J., Porter, J.H.	2017	Seagrass blue carbon spatial patterns at the meadow-scale	Peer-reviewed article	*PLoS One*	The United States	Seagrasses	X				X
Pedersen, M.F., Filbee-Dexter, K., Norderhaug, K.M., Fredriksen, S., Frisk, N.L., Fagerli, C.W., Wernberg, T.	2019	Detrital carbon production and export in high latitude kelp forests	Peer-reviewed article	*Oecologia*	Norway	Kelp (*Laminaria hyperborean*)	X				
Pessarrodona, A., Moore, P.J., Sayer, M.D.J., Smale, D.A.	2018	Carbon assimilation and transfer through kelp forests in the NE Atlantic is diminished under a warmer ocean climate	Peer-reviewed article	*Global Change Biology*	NE Atlantic Ocean	Kelp (*Laminaria hyperborean*)	X				X
Krause-Jensen, D., Lavery, P., Serrano, O., Marbá,N., Masque, P., Duarte, C.M.	2018	Sequestration of macroalgal carbon: the elephant in the Blue Carbon room	Peer-reviewed article	*Biology Letters*	Global	Macroalgae	X		X	X	X

(Continued)

TABLE 14.1 Results of the systematic quantitative literature review *(Continued)*

Author	*Year*	*Title*	*Type*	*Journal*	*Study location*	*Species*	*Scientific*			*Regulatory*	
							Technical	*Social*	*Economic*	*Policy*	*Management*
Smale, D. A., Moore, P. J., Queirós, A. M., Higgs, S., & Burrows, M. T.	2018	Appreciating interconnectivity between habitats is key to blue carbon management	Peer-reviewed article	*Frontiers in Ecology and the Environment*	Global	Mangroves, seagrasses, macroalgae	X			X	X
Smale, D. A., Burrows, M. T., Moore, P., O'Connor, N., & Hawkins, S. J.	2013	Threats and knowledge gaps for ecosystem services provided by kelp forests: a northeast Atlantic perspective	Peer-reviewed article	*Ecology and Evolution*	NE Atlantic Ocean	Kelp	X	X	X	X	X
Sondak, C.F.A., Chung, I.K.	2015	Potential Blue Carbon from Coastal Ecosystems in the Republic of Korea	Peer-reviewed article	*Ocean Science*	South Korea	Mangroves, seagrasses, tidal marshes, seaweeds, macroalgae, kelp	X				

(Continued)

TABLE 14.1 Results of the systematic quantitative literature review *(Continued)*

							Scientific			Regulatory	
Author	*Year*	*Title*	*Type*	*Journal*	*Study location*	*Species*	*Technical*	*Social*	*Economic*	*Policy*	*Management*
Sondak, C.F.A., Ang Jr., P.O., Beardall, J., Bellgrove, A., Boo, S.M., Gerung, G.S., Hepburn, C.D., Hong, D.D., Hu, Z., Kawai, H., Largo, D., Lee, J.A., Lim, P., Mayakun, J., Nelson, W., Oak, J.H., Phang, S., Sahoo, D., Peerapornpis, Y., Yang, Y., Chung, I.K.	2016	Carbon dioxide mitigation potential of seaweed aquaculture beds (SABs)	Peer-reviewed article	*Journal of Applied Phycology*	Asian-Pacific region	Mangroves, seagrasses, salt marshes, seaweeds, macroalgae	X		X		
Thompson, B. S., Clubbe, C. P., Primavera, J. H., Curnick, D., & Koldewey, H.	2014	Locally assessing the economic viability of blue carbon: A case study from Panay Island, the Philippines	Peer-reviewed article	*Journal of Applied Phycology*	Philippines	Mangroves	X	X	X	X	X

(Continued)

TABLE 14.1 Results of the systematic quantitative literature review *(Continued)*

							Scientific			*Regulatory*	
Author	*Year*	*Title*	*Type*	*Journal*	*Study location*	*Species*	*Technical*	*Social*	*Economic*	*Policy*	*Management*
Trevathan-Tackett, S. M., Kelleway, J., Macreadie, P. I., Beardall, J., Ralph, P., & Bellgrove, A.	2015	Comparison of marine macrophytes for their contributions to blue carbon sequestration	Peer-reviewed article	*Ecology*	Australia	Mangroves, seagrasses, salt marshes, macroalgae, kelp	X			X	
Vanderklift, M. A., Marcos-Martinez, R., Butler, J. R., Coleman, M., Lawrence, A., Prislan, H., ... & Thomas, S.	2019	Constraints and opportunities for market-based finance for the restoration and protection of blue carbon ecosystems	Peer-reviewed article	*Marine Policy*	Global	Mangroves, seagrasses, tidal marshes	X	X	X	X	X
Wylie, L., Sutton-Grier, A.E., Moore, A.	2016	Keys to successful blue carbon projects: Lessons learned from global case studies	Peer-reviewed article	*Marine Policy*	Kenya, Vietnam, India/ Bangladesh, Madagascar	Mangroves, seagrasses, salt marshes	X	X	X	X	
Zarate-Barrera, T.G., Maldonado, J.H.	2015	Valuing Blue Carbon: Carbon Sequestration Benefits Provided by the Marine Protected Areas in Colombia	Peer-reviewed article	*PLoS One*	Colombia	Mangroves, seagrasses	X	X	X	X	

(Continued)

TABLE 14.1 Results of the systematic quantitative literature review *(Continued)*

Author	*Year*	*Title*	*Type*	*Journal*	*Study location*	*Species*	*Scientific*			*Regulatory*	
							Technical	*Social*	*Economic*	*Policy*	*Management*
Bayley, D.T.I., Marengo, I., Pelembe, T.	2017	Giant kelp "Blue carbon" storage and sequestration value in the Falkland Islands	Research Report		Falkland Islands	*Macrocystis* kelp forests	X				
Kuwae, T., Hori, M.	2019	The Future of Blue Carbon: Addressing Global Environmental Issues	Book chapter		Global	Mangroves, seagrasses, salt marshes, tidal marshes, macroalgae	X	X	X	X	X

FIGURE 14.1 Treemap displaying the numbers of references for terms found in articles published from 2013 to 2019

the original list of the top 20 references, this term was included for comparison. The words "Seagrass," "Mangrove," "Seaweed," "Kelp," and "Nitrogen" were also included from the list of words in the WFQ in Figure 14.2. The term "Salt marsh" was also added for comparison because it is considered a traditional BC ecosystem. "Tidal marsh" was excluded from the Miscellaneous section because it had fewer than 2% of the total references. These six subthemes were then subjected to individual TSQs that accounted for singular, plural, and abbreviated

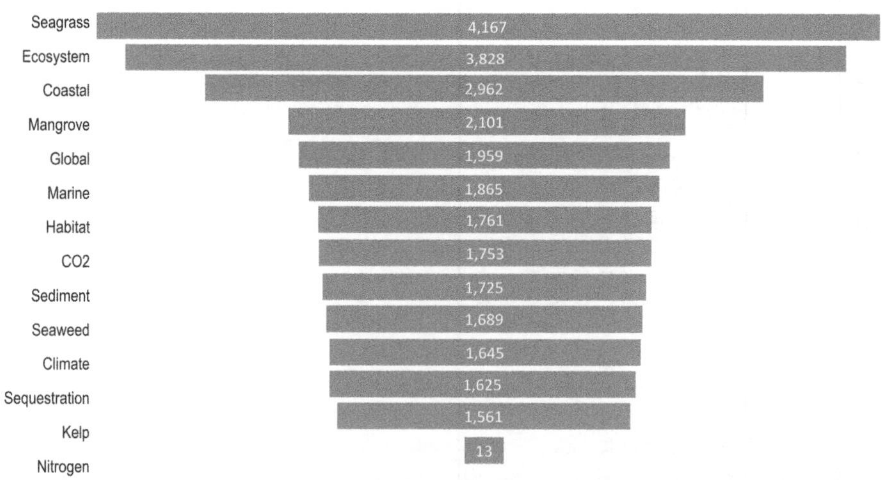

FIGURE 14.2 Word Frequency Query results displaying the number of references for each word found in articles published from 2013 to 2019

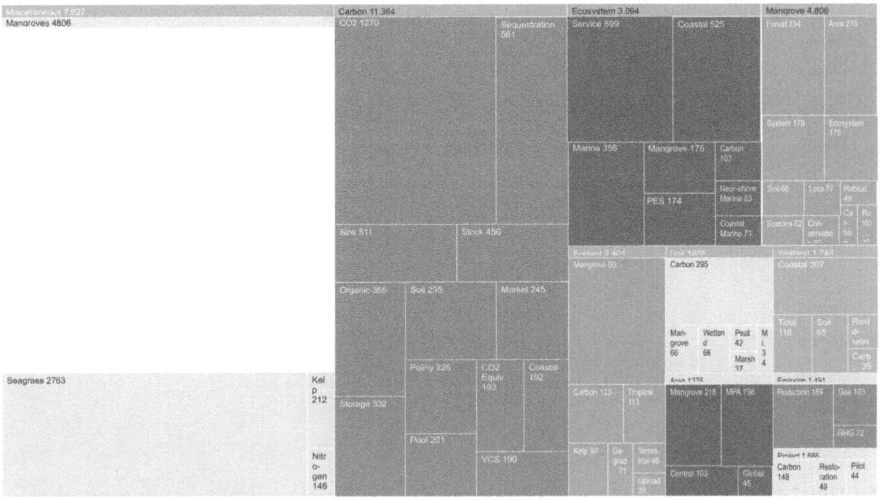

FIGURE 14.3 Treemap displaying the numbers of references for terms found in articles published from 2009 to 2013

references. The summed results were displayed in the "Miscellaneous" section of the treemap in Figure 14.1.

Because Thomas's original meta-analysis was conducted with a different qualitative analysis software called Leximancer, a second thematic auto-coding was performed on his original 46-article set using the NVivo automated insights auto-code feature. This thematic analysis generated 9 overarching themes with 2,251 subthemes. Within each overarching theme, the subthemes were manually aggregated into categories with 25 references or greater per subtheme. After the manual aggregation, all the overarching themes and subthemes with 25 references or greater were each individually subjected to TSQs in which references to each word or phrase were summed across the 46 articles. The TSQs accounted for "stemmed words," and the singular, plural, and abbreviated forms of each reference were totalled. The results were displayed in the treemap in Figure 14.3.

A "Miscellaneous" overarching theme was also added to this analysis. A WFQ was performed on the 46-article set, and the top 20 references across the entire body of literature were displayed in Figure 14.4. References to "Blue," "Carbon," and any other stop words that were deemed as clutter were removed. Stemmed words were included to account for different forms of the same word. Despite not falling in the top 20 references, the term "Nitrogen" was again included in the WFQ for comparison. The terms "Mangrove" and "Nitrogen" were selected from the list of words in the WFQ in Figure 14.4. The terms "Seagrass," "Salt marsh," and "Kelp" were also included as additional subthemes of interest because they are considered both traditional and non-traditional BC ecosystems. "Tidal marsh" was excluded from the Miscellaneous section because it had fewer than 2% of the total references. These four subthemes were then

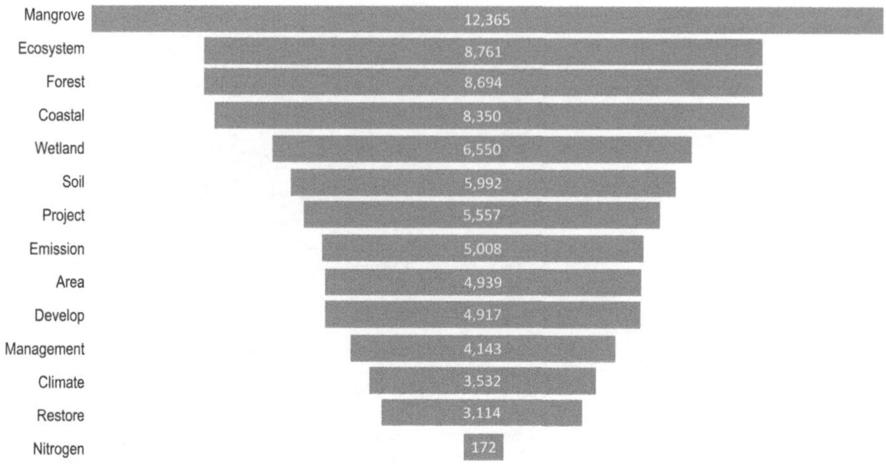

FIGURE 14.4 Word Frequency Query results for articles displaying the number of references for each word found in articles published from 2009 to 2013

subjected to individual TSQs that accounted for singular, plural, and abbreviated references. The results were displayed in the "Miscellaneous" section of the treemap in Figure 14.3.

Finally, a third auto-code analysis was conducted on the combination of Thomas's 46-article set and our 38-article set. The automated insights auto-code feature generated 6 overarching themes and 2,539 subthemes that were manually aggregated into categories with 25 or greater references each. All the overarching themes and subthemes with 25 references or greater were each individually subjected to TSQs that summed the total number of references across the 84-article set. The TSQs included "stemmed words," and the singular, plural, and abbreviated forms of each overarching theme and subtheme were totalled. The summed results were displayed in a treemap in Figure 14.5.

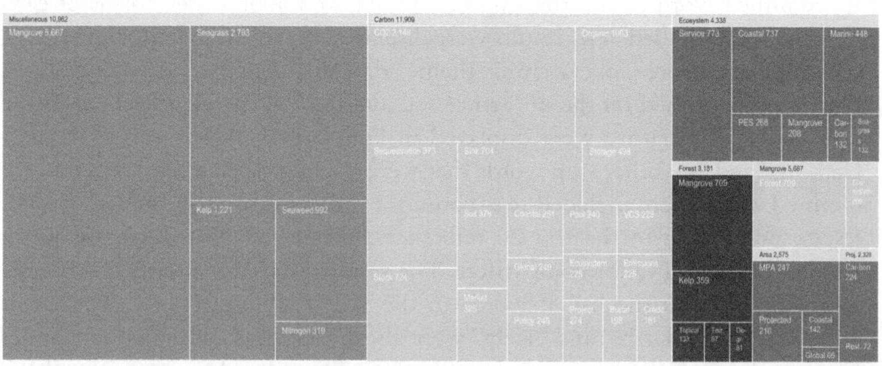

FIGURE 14.5 Word Frequency Query results for articles displaying the number of references for each word found in articles published from 2009 to 2019

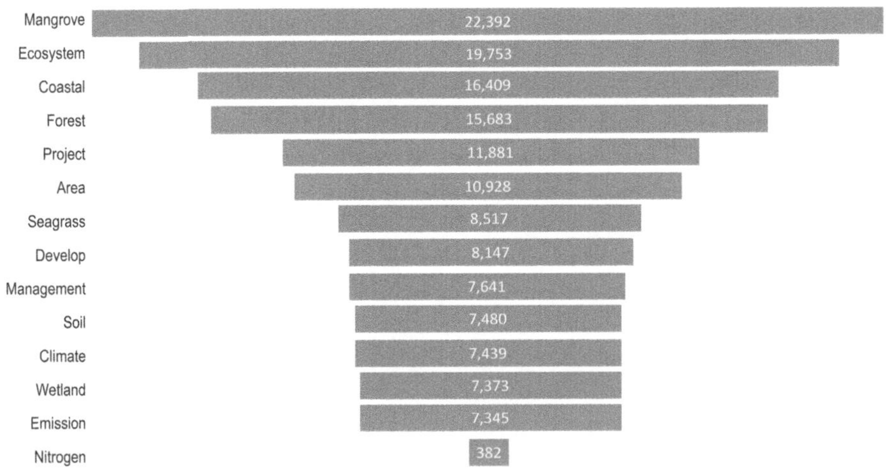

FIGURE 14.6 Word Frequency Query results displaying the number of references for each word found in articles published from 2009 to 2019

A "Miscellaneous" overarching theme was added to this third analysis. A WFQ was performed on the 84-article set, and the top 20 references in the total body of literature were displayed in Figure 14.6. References to "Blue," "Carbon," and any other stop words that were deemed as clutter were removed. Stemmed words were included to account for different variations of the same word. The term "Nitrogen" was included in the WFQ for comparison even though it did not appear on the list of top 20 references. The words "Mangrove," "Seagrass," and "Nitrogen" were included from the list of words in the WFQ in Figure 14.6. The terms "Kelp," "Seaweed," and "Tidal marsh" were included as additional subthemes of interest because they are considered traditional and non-traditional BC ecosystems. "Tidal marsh" was excluded from the Miscellaneous section because it had fewer than 2% of the total references. These five subthemes were then subjected to individual TSQs that accounted for singular, plural, and abbreviated references. The summed results were displayed in the "Miscellaneous" section of the treemap in Figure 14.5.

Because the term "Blue Carbon" was a central concept in all the documents, this theme was removed from all three analyses to prevent a skewing of the results. Stop words that did not possess relevant meaning were also extricated to provide additional elucidation and clarity. Furthermore, if a subtheme had 25 or greater references but also comprised less than 2% of the total number of references in the overarching theme, it was not displayed in the corresponding treemap.

14.3 Results

Of the 38 articles selected for this investigation, 36 were peer-reviewed papers in academic journals, 1 was a report generated by a research institution, and 1 was a chapter in a published book. The 36 peer-reviewed papers were published in

23 different journals. The publication window spanned from July 2013 through December 2019. Of the four articles published in 2013, only one article overlapped with Thomas's literature search window, which concluded in July 2013. However, this chapter was not included in Thomas's meta-analysis, so no actual overlap occurred. The remaining three articles in this investigation from 2013 were published in August and September of that year. In the subsequent five years, five articles were published in 2014 and 2015, respectively, seven articles were published in 2016, nine articles were published in 2017, two articles were published in 2018, and six articles were published in 2019.

Fifteen of the study locations were conducted on a global scale, and 13 studies occurred in individual countries, with Australia and the United States tying for the greatest number of study locations. The remaining countries of origin included study sites in Africa, Asia, Europe, the Middle East, and South America. Furthermore, four studies were conducted across broad regions including Europe, the Northeast Atlantic Ocean, and the Asian Pacific region. Among the species examined at these locations, seagrasses were the most prevalently studied species, followed closely by mangroves. Salt marshes, seaweeds, kelp, and macroalgae were all studied at about the same frequency, which was approximately half that of mangroves. Tidal marshes were the least studied habitat. After closely scrutinising the entire body of texts, each article was also categorised based upon the following themes: science/technology, social, economic, regulatory/policy, and management. Thirty-six articles were scientific or technical in nature. Seventeen papers dealt with regulatory/policy and management topics, respectively. Eleven articles contained economic material, and only seven papers included social elements.

Of the treemaps generated from the NVivo qualitative analysis, the overarching theme containing the greatest number of references was "Carbon" across all three analyses. The 84-article set had the most references to "Carbon" (11,909) followed by the 46-article set (11,364) and the 38-article set (4,991). Within the "Carbon" overarching theme, "Carbon dioxide" had the greatest number of references across all three analyses. This subtheme comprised 13% of the "Carbon" overarching themes in the 38-article set, 11% in the 46-article set, and 18% in the 84-article set. "Organic carbon" (8%) and "Carbon sequestration" (5%) had the second greatest number of references in the 38- and 46-article sets, and "Organic carbon" (8%) and "Carbon sequestration" (8%) were tied for the second greatest number of references in the 84-article set. "Carbon sequestration" (5%) garnered the third highest number of references in the 38-article set while "Carbon sink" (4%) and "Carbon stock" (4%) were tied for the third highest number of references in the 46-article set. "Carbon stock" (6%) and "Carbon sink" (6%) were also tied for the third highest number of references in the 84-article set.

The overarching theme with the second greatest number of references shared across all three bodies of literature was the "Miscellaneous" overarching theme, which had the greatest number of references in the 84-article set (11,763), followed by the 46-article set (7,097) and the 38-article set (4,656).

Within the "Miscellaneous" category, "Seagrass" (31%), "Mangrove" (68%), and "Mangrove" (48%) were the subthemes with the greatest number of references in the 38-, 46-, and 84-article sets, respectively. "Kelp" (22%), "Seagrass" (19%), and "Seagrass" (23%) were the subthemes with the second greatest number of references across all three sets, respectively. "Seaweed" (21%), "Salt marsh" (8%), and "Kelp" (10%) were the subthemes with the third greatest number of references in the 38, 46, and 84-article sets.

The overarching theme with the third greatest number of references across all three article sets was "Ecosystem," which had the greatest number of references in the 84-article set (4,295), followed by the 46-article set (3,051) and the 38-article set (1,244). Within this overarching theme, "Coastal ecosystem" (17%), "Ecosystem service" (20%), and "Ecosystem service" (18%) were the subthemes with the greatest number of references in the 38-, 46-, and 84-article sets. "Ecosystem service" (14%), "Coastal ecosystem" (17%), and "Coastal ecosystem" (17%) were the subthemes with the second greatest number of references across the three article sets. "Payments for ecosystem services," "Marine ecosystem" (12%), and "Marine ecosystem" (10%) were the subthemes with the third greatest number of references in the 38-, 46-, and 84-article sets.

Of the remaining overarching themes in the 38-article set, "Production" (886), "Rate" (721), and "Area" (699) had the fewest number of references. In the 46-article set, "Mangrove" (4,806) was the overarching theme with second greatest number of references overall, but this theme did not appear in all three article sets. The overarching themes with the fewest number of references in the 46-article set were "Forest" (2,401), "Area" (1,876), "Wetland" (1,748), "Project" (1,686), "Soil" (1,605), and "Emission" (1,535). The overarching theme "Mangrove" had the third greatest number of references within the 84-article set (5,667) but did not appear in all three article sets. The overarching themes with the fewest number of references in this article set were "Forest" (3,181), "Area" (2,575), and "Project" (2,328).

14.4 Discussion

These results suggest the growing importance of macroalgae such as seaweeds and kelps in the BC discourse. However, a debate has emerged in the literature about whether macroalgae should be considered true BC ecosystems. The overwhelming presence of the term "Carbon sequestration" in all three NVivo analyses reinforces the idea that a BC ecosystem's ability to sequester carbon lies central to the definition of BC. In comparison, relatively little attention has been paid to carbon offsetting, which did not appear as a subtheme in any of the treemaps. PES also garnered relatively little attention, as it was mentioned fewer times in the 38-article set than in the 46-article set. This decline becomes relevant when considering that both bodies of literature focused solely on the potential for monetising carbon sequestration through PES while neglecting BC ecosystems' nitrogen sequestration capacity.

14.4.1 Macroalgae as Non-traditional BC Ecosystems

References to macroalgae were scant in the 46-article set, with kelp garnering only 2.7% of the total references in the corresponding Miscellaneous section. In comparison, traditional BC ecosystems such as mangroves and seagrasses comprised 61% and 35% of this section, respectively. The term "Seaweed" was excluded entirely because it was referenced only ten total times, falling well short of the 2% cut-off for inclusion. However, references to seaweed and kelp rose to 22% and 23%, respectively, in the Miscellaneous section of the 38-article set and comprised 9% and 11% of this section in the combined 84-article set.

This rise in references to macroalgae in the BC literature can be attributed to the recent debate that erupted amongst scholars who initially claimed that because macroalgae grow predominantly on rocky substrates and do not develop widespread rooting systems for capturing sediments and detritus, they cannot directly sequester carbon for long-term periods in the surrounding soil like traditional BC ecosystems such as mangroves, seagrasses, and tidal marshes (Ahmed et al., 2016; Bayley et al., 2017; Chung et al., 2017; Hill et al., 2015; Howard et al., 2017; Krause-Jensen and Duarte, 2016; Trevathan-Tackett et al., 2015). Carbon sequestration emerged as the key defining theme of BC in the NVivo analyses across all three article sets.

Within the 84-article set, "Carbon sequestration" was tied with "Organic carbon" for the second greatest number of references by percentage in the Carbon section and was only surpassed in number of references by the ubiquitous term "Carbon dioxide." In addition to the demonstrated centrality of carbon sequestration to the definition of BC, two other subthemes emerged as relevant contenders. The prevalence of the terms "Carbon stock" and "Carbon sink," which were tied for the third greatest number of references by percentage in the 84-article set, suggests that these concepts should also be considered in the working definition. Because macroalgae contribute substantially to global carbon stocks (Howard et al., 2017) while functioning as massive carbon sinks (Bayley et al., 2017; Chung et al., 2017; Hill et al., 2015; Krause-Jensen et al., 2018; Sondak and Chung, 2015; Sondak et al., 2017) and carbon donors (Ahmed et al., 2016; Hill et al., 2015; Krause-Jensen and Duarte, 2016; Pedersen, et al., 2020; Smale et al., 2018; Trevathan-Tackett et al., 2015), macroalgae clearly contribute to the overall BC picture, albeit in different manners than traditional BC ecosystems. Despite the push to discount macroalgae as true BC ecosystems, this chapter argues that seaweeds and kelps should be considered non-traditional contributors to BC because they function as natural and formidable carbon sinks and donors to other traditional BC sources.

Researchers have confirmed that kelp contributes substantial quantities of allochthonous carbon, which refers to organic matter that is transported far away from its originating position and is sequestered belowground by other BC ecosystems (Krause-Jensen et al., 2018; Smale et al., 2018; Sondak et al., 2017).

However, current carbon crediting mechanism currently excludes allochthonous carbon from their accounting schemes (Krause-Jensen et al., 2018; Smale et al., 2018). Permanence is currently a requirement of the Verified Carbon Standard, which is the most utilised verification standard for carbon mitigation projects under the International Panel on Climate Change (Krause-Jensen et al., 2018). Labile BC sources such as macroalgae have been considered less suitable to long-term sequestration (Pedersen et al., 2020) even though when their allochthonous biomass is exported to the open ocean and sequestered in deep-sea sediments, macroalgae donate an estimated mass of organic carbon that is comparable to the carbon sequestered by all other BC habitats combined (Froehlich et al., 2019; Krause-Jensen et al., 2018). Therefore, this chapter contends that the definition of BC should be expanded to include non-traditional sources of BC such as macroalgae.

14.4.2 Dearth in the Discussion of Carbon Offsetting

Although the term "Carbon offset" was mentioned in both the 38- and 46-article sets, it contained less than 2% of the total references in the respective Carbon sections and was thus excluded from the corresponding treemaps. This suggests that the past and current BC literature has been more focused on the sequestration potential of traditional and non-traditional ecosystems while neglecting how to capitalise on this potential through carbon offset markets. Traditionally, developing carbon offset credits has encompassed a top-down approach that focuses on establishing international markets first before creating credit markets at national and local levels of government (Kuwae and Hori, 2019). Furthermore, for local BC credit schemes to succeed, independent measurement, reporting, and verification methods need to be developed (Kuwae and Hori, 2019).

The development of international regulatory frameworks has generated a rapid expansion in the global carbon market, which increased from 11 billion USD to 141.9 billion USD from 2005 to 2010 (Linacre et al., 2011). Currently, the forest carbon market is divided between compliance schemes, which are created and regulated by national and international agreements, and voluntary mechanisms, which are dominated by the investments of companies and individuals (Locatelli et al., 2014). Researchers have argued that mangrove forests have the potential to generate carbon offset credits in a voluntary market scheme (Locatelli et al., 2014) and have demonstrated the potential for generating carbon offset credits in seagrass meadows (Oreska et al., 2017, 2018).

However, the inclusion of other non-traditional BC ecosystems such as macroalgae into existing carbon offsetting schemes remains murky. Part of the difficulty with monetising the sequestration potential of both traditional and non-traditional BC ecosystems through current carbon offsetting mechanisms occurs because the United Nations Framework Convention on Climate Change applies the terrestrial time scale of at least 50 years required for afforestation to the sequestration timeline of oceanic BC (Chung et al., 2017). Because non-traditional BC ecosystems do not contribute to carbon sequestration in the same manner as terrestrial forests, macroalgae will continue to be excluded by international regulatory bodies unless a separate set of regulations are developed that encompass both traditional and non-traditional BC ecosystems.

14.4.3 The Importance of Payment for Ecosystem Services

Although references to PES occur in both the 38- and 46-article sets, the total number of references to PES declined slightly from Thomas's meta-analysis to ours. PES further enriches the idea of carbon offsetting by creating financial incentives for conservation and generating revenue through the monetisation of services that ecosystems naturally provide without cost to humanity (Thompson et al., 2014). Inherent to the idea of PES is the plurality of services that can be monetised. Carbon sequestration represents only one of the potentially monetisable ecosystem services that traditional and non-traditional BC ecosystems provide.

While some researchers acknowledged that in addition to carbon dioxide capture, seaweed provides other vital ecosystem services such as coastal protection, improving water quality, and nutrient cycling (Chung et al., 2017), regulatory frameworks do not yet exist to financially capitalise on these services. In fact, of the current PES schemes that exist, most operate under international policies such as Reduced Emissions from Deforestation and Degradation programs, which focus predominantly on terrestrial habitats and typically ignore BC projects (Locatelli et al., 2014).]. Yet, the plurality of services that both traditional and non-traditional BC ecosystems offer could increase their attractiveness to investors by differentiating offsets and monetising their co-benefits (Vanderklift et al., 2019).

14.4.4 The Role of Blue Nitrogen in PES

The treemaps from all three analyses demonstrate that the number of references to "Nitrogen" were negligible in comparison to "Carbon," "Carbon dioxide," and "Carbon sequestration." In the 84-article set, "Nitrogen" was referenced only 319 times in comparison to "Carbon," which was mentioned nearly 12,000 times across the entire body of literature. Because carbon storage is just one of a plethora of ecosystem services that traditional and non-traditional BC ecosystems offer to society (Bayley et al., 2017), it remains important not to view this service in isolation but rather regard it as a part of an enmeshed system that is collectively greater than the sum of its parts. An additional PES opportunity that was not considered by either the 38- or 46-article sets is the potential for "Blue Nitrogen," which the authors define as the ability of macroalgae such as kelp and seaweed to extract nitrogen from the surrounding waters. Through the process of bio extraction, seaweeds have been used to remove and sequester excess nutrients such as nitrogen from urban coastal waters (Kim et al., 2014, 2015). Yet, little emphasis has been placed on this ecosystem service as a potential monetisable PES scheme in either body of literature that was examined.

14.5 Conclusion and Implications for the Blue Economy

This meta-analysis has argued for the inclusion of macroalgae in the BC discourse while highlighting the need to better understand how to maximise the carbon offset-generating potential of traditional and non-traditional BC ecosystems through the bundling BC and Blue Nitrogen in a cohesive PES framework. To accomplish these aims, further research is needed to address the double counting of macroalgae-derived allochthonous carbon deposited in traditional BC ecosystems. Because marine environments are highly connected and subject to large inputs of allochthonous carbon, future studies are needed to advance accounting techniques that can identify the sources of BC in traditional BC ecosystems to avoid the double counting of carbon from macroalgal donor sites (Krause-Jensen et al., 2018). The issue of carbon leakage in BC ecosystems also requires further investigation due to the complexities of monitoring leakage and the paucity of thorough assessments in the context of BC (Locatelli et al., 2014). Leakage issues also become prominent when considering the present and future threats that climate change poses by disturbing BC ecosystems.

Although the complexities of carbon offset crediting have been framed by a top-down approach, shifting to a bottom-up approach in which new markets are first established at the level of local government and subsequently reinforced by national and international organisations may facilitate the development of BC credit schemes (Kuwae and Hori, 2019). The maturation of BC carbon crediting is a necessary first step towards developing a more wholistic approach to PES. With such great focus currently being placed on the carbon sequestration capacity of BC ecosystems, the current body of literature has yet to analyse the offset-generating potential of Blue Nitrogen. Future studies are needed to explore the incorporation of Blue Nitrogen into existing carbon offset markets, regulatory frameworks, and PES schemes. In closing, BC and Blue Nitrogen markets have the potential to become significant sectors of the burgeoning Blue Economy and thus warrant further analysis. Further research should focus on assessing if these markets have the ability to contribute to increased livelihood opportunities in the developing world. The potential for this has been argued for by Steven et al. (2019) in their editorial of a themed section in the *Journal of the Indian Ocean Region* (2019, Volume 19, pp. 123–234), which may well prove to be an excellent starting point for this type of work.

Acknowledgements

The authors would like to thank Dr. Miriah Kelly for her assistance with the NVivo 12 Plus software package. This research was supported by an award from the New England Saint Lawrence Valley Geographical Society.

References

Ahmed, N., Bunting, S. W., Glaser, M., Flaherty, M. S., & Diana, J. S. (2016). Can greening of aquaculture sequester blue carbon?. *Ambio, 46*(4), 468–477. doi:10.1007/s13280-016-0849-7.

Bayley, D. T. I., Marengo, I., & Pelembe, T. (2017). *Giant kelp 'Blue carbon' storage and sequestration value in the Falkland Islands.* doi:10.13140/RG.2.2.31988.24966.

Chung, I. K., Sondak, C. F., & Beardall, J. (2017). The future of seaweed aquaculture in a rapidly changing world. *European Journal of Phycology, 52*(4), 495–505. doi:10.1080/09670262.2017.1359678.

Ferdouse, F., Holdt, S. L., Smith, R., Murúa, P., & Yang, Z. (2018). The global status of seaweed production, trade and utilization. *Globefish Research Programme, 124,* I.

Froehlich, H. E., Afflerbach, J. C., Frazier, M., & Halpern, B. S. (2019). Blue growth potential to mitigate climate change through seaweed offsetting. *Current Biology, 29*(18), 3087–3093. doi:10.1016/j.cub.2019.07.041.

Hill, R., Bellgrove, A., Macreadie, P. I., Petrou, K., Beardall, J., Steven, A., & Ralph, P. J. (2015). Can macroalgae contribute to blue carbon? An Australian perspective. *Limnology and Oceanography, 60*(5), 1689–1706. doi:10.1002/lno.10128.

Howard, J., Sutton-Grier, A., Herr, D., Kleypas, J., Landis, E., Mcleod, E., ... & Simpson, S. (2017). Clarifying the role of coastal and marine systems in climate mitigation. *Frontiers in Ecology and the Environment, 15*(1), 42–50. doi:10.1002/fee.1451.

Kim, J. K., Kraemer, G. P., & Yarish, C. (2014). Field scale evaluation of seaweed aquaculture as a nutrient bioextraction strategy in Long Island Sound and the Bronx River Estuary. *Aquaculture, 433,* 148–156. doi:10.1016/j.aquaculture.2014.05.034.

Kim, J. K., Kraemer, G. P., & Yarish, C. (2015). Use of sugar kelp aquaculture in Long Island Sound and the Bronx River Estuary for nutrient extraction. *Marine Ecology Progress Series, 531,* 155–166. doi:10.3354/meps11331.

Krause-Jensen, D., & Duarte, C. M. (2016). Substantial role of macroalgae in marine carbon sequestration. *Nature Geoscience, 9*(10), 737–742. doi:10.1038/ngeo2790.

Krause-Jensen, D., Lavery, P., Serrano, O., Marba, N., Masque, P., & Duarte, C. M. (2018). Sequestration of macroalgal carbon: the elephant in the Blue Carbon room. *Biology Letters, 14*(6), 20180236. doi:10.1098/rsbl.2018.0236.

Kuwae, T., & Hori, M. (2019). The future of blue carbon: addressing global environmental issues. *Blue Carbon in Shallow Coastal Ecosystems* (pp. 347–373). Springer, Singapore. doi:10.1007/978-981-13-1295-3_13.

Linacre, N., Kossoy, A., & Ambrosi, P. (2011). *State and trends of the carbon market 2011.*

Locatelli, T., Binet, T., Kairo, J. G., King, L., Madden, S., Patenaude, G., ... & Huxham, M. (2014). Turning the tide: how blue carbon and payments for ecosystem services (PES) might help save mangrove forests. *Ambio, 43*(8), 981–995. doi:10.1007/s13280-014-0530-y.

Lovelock, C. E., & Duarte, C. M. (2019). Dimensions of blue carbon and emerging perspectives. *Biology Letters, 15*(3), 20180781. doi:10.1098/rsbl.2018.0781.

McHugh, D. J. (2003). *A guide to the seaweed industry.*

Mcleod, E., Chmura, G. L., Bouillon, S., Salm, R., Björk, M., Duarte, C. M., ... & Silliman, B. R. (2011). A blueprint for blue carbon: toward an improved understanding of the role of vegetated coastal habitats in sequestering CO_2. *Frontiers in Ecology and the Environment, 9*(10), 552–560. doi:10.1890/110004.

NVivo. (2020). Retrieved on September 16, 2020 from https://www.qsrinternational.com/nvivo-qualitative-data-analysis-software/home

Oreska, M. P., McGlathery, K. J., & Porter, J. H. (2017). Seagrass blue carbon spatial patterns at the meadow-scale. *PLoS One, 12*(4), e0176630. doi:10.1371/journal.pone.0176630.

Oreska, M. P., Wilkinson, G. M., McGlathery, K. J., Bost, M., & McKee, B. A. (2018). Non-seagrass carbon contributions to seagrass sediment blue carbon. *Limnology and Oceanography, 63*(S1), S3–S18. doi:10.1002/lno.10718.

Pedersen, M. F., Filbee-Dexter, K., Norderhaug, K. M., Fredriksen, S., Frisk, N. L., & Wernberg, T. (2020). Detrital carbon production and export in high latitude kelp forests. *Oecologia, 192*(1), 227–239. doi:10.1007/s00442-019-04573-z.

Smale, D. A., Moore, P. J., Queirós, A. M., Higgs, S., & Burrows, M. T. (2018). Appreciating interconnectivity between habitats is key to blue carbon management. *Frontiers in Ecology and the Environment, 16*(2), 71–73. doi:10.1002/fee.1765.

Sondak, C. F., & Chung, I. K. (2015). Potential blue carbon from coastal ecosystems in the Republic of Korea. *Ocean Science Journal, 50*(1), 1–8. doi:10.1007/s12601-015-0001-9.

Sondak, C. F., Ang, P. O., Beardall, J., Bellgrove, A., Boo, S. M., Gerung, G. S., … & Largo, D. (2017). Carbon dioxide mitigation potential of seaweed aquaculture beds (SABs). *Journal of Applied Phycology, 29*(5), 2363–2373. doi:10.1007/s10811-016-1022-1.

Steven, A. D., Vanderklift, M. A., & Bohler-Muller, N. (2019). A new narrative for the Blue Economy and blue carbon. *Journal of the Indian Ocean Region, 15*(2), 123–128.

Thomas, S. (2014). Blue carbon: Knowledge gaps, critical issues, and novel approaches. *Ecological Economics, 107*, 22–38. doi:10.1016/j.ecolecon.2014.07.028.

Thompson, B. S., Clubbe, C. P., Primavera, J. H., Curnick, D., & Koldewey, H. J. (2014). Locally assessing the economic viability of blue carbon: a case study from Panay Island, the Philippines. *Ecosystem Services, 8*, 128–140. doi:10.1016/j.ecoser.2014.03.004.

Trevathan-Tackett, S. M., Kelleway, J., Macreadie, P. I., Beardall, J., Ralph, P., & Bellgrove, A. (2015). Comparison of marine macrophytes for their contributions to blue carbon sequestration. *Ecology, 96*(11), 3043–3057. doi:10.1890/15-0149.1.

Vanderklift, M. A., Marcos-Martinez, R., Butler, J. R., Coleman, M., Lawrence, A., Prislan, H., … & Thomas, S. (2019). Constraints and opportunities for market-based finance for the restoration and protection of blue carbon ecosystems. *Marine Policy, 107*, 103429. doi:10.1016/j.marpol.2019.02.001.

Appendix A

References to the 38 articles selected for this investigation.

Ahmed, N., Bunting, S. W., Glaser, M., Flaherty, M. S., & Diana, J. S. (2016). Can greening of aquaculture sequester blue carbon?. *Ambio, 46*(4), 468–477. doi:10.1007/s13280-016-0849-7.

Alongi, D. M., Murdiyarso, D., Fourqurean, J. W., Kauffman, J. B., Hutahaean, A., Crooks, S., … & Pidgeon, E. (2016). Indonesia's blue carbon: a globally significant and vulnerable sink for seagrass and mangrove carbon. *Wetlands Ecology and Management, 24*(1), 3–13. doi:10.1007/s11273-015-9446-y.

Bayley, D. T. I., Marengo, I., & Pelembe, T. (2017). *Giant kelp 'Blue carbon' storage and sequestration value in the Falkland Islands.* 10.13140/RG.2.2.31988.24966

Campbell, I., Macleod, A., Sahlmann, C., Neves, L., Funderud, J., Øverland, M., … & Stanley, M. (2019). The environmental risks associated with the development of seaweed farming in Europe-prioritizing key knowledge gaps. *Frontiers in Marine Science, 6*, 107. doi:10.3389/fmars.2019.00107.

Campbell, J. E., Lacey, E. A., Decker, R. A., Crooks, S., & Fourqurean, J. W. (2015). Carbon storage in seagrass beds of Abu Dhabi, United Arab Emirates. *Estuaries and Coasts*, *38*(1), 242–251. doi:10.1007/s12237-014-9802-9.

Chung, I. K., Sondak, C. F., & Beardall, J. (2017). The future of seaweed aquaculture in a rapidly changing world. *European Journal of Phycology*, *52*(4), 495–505. doi:10.1080/09670262.2017.1359678.

Duarte, C. M. (2017). Reviews and syntheses: hidden forests, the role of vegetated coastal habitats in the ocean carbon budget. *Biogeosciences*, 14(2), 301–310. doi:10.5194/bg-14-301-2017.

Duarte, C. M., Wu, J., Xiao, X., Bruhn, A., & Krause-Jensen, D. (2017). Can seaweed farming play a role in climate change mitigation and adaptation?. *Frontiers in Marine Science*, *4*, 100. doi:10.3389/fmars.2017.00100.

Erlania, E., & Radiarta, I. N. (2014). Management of sustainable seaweed (*Kappaphycus alvarezii*) aquaculture in the context of climate change mitigation. *Indonesian Aquaculture Journal*, *9*(1), 65–72. doi:10.15578/iaj.9.1.2014.65-72.

Froehlich, H. E., Afflerbach, J. C., Frazier, M., & Halpern, B. S. (2019). Blue growth potential to mitigate climate change through seaweed offsetting. *Current Biology*, *29*(18), 3087–3093. doi:10.1016/j.cub.2019.07.041.

Greiner, J. T., McGlathery, K. J., Gunnell, J., & McKee, B. A. (2013). Seagrass restoration enhances "blue carbon" sequestration in coastal waters. *PLoS One*, *8*(8), e72469. doi:10.18130/v3f672.

Herr, D., von Unger, M., Laffoley, D., & McGivern, A. (2017). Pathways for implementation of blue carbon initiatives. *Aquatic Conservation: Marine and Freshwater Ecosystems*, *27*, 116–129. doi:10.1002/aqc.2793.

Hill, R., Bellgrove, A., Macreadie, P. I., Petrou, K., Beardall, J., Steven, A., & Ralph, P. J. (2015). Can macroalgae contribute to blue carbon? An Australian perspective. *Limnology and Oceanography*, *60*(5), 1689–1706. doi:10.1002/lno.10128.

Howard, J., Sutton-Grier, A., Herr, D., Kleypas, J., Landis, E., Mcleod, E., ... & Simpson, S. (2017). Clarifying the role of coastal and marine systems in climate mitigation. *Frontiers in Ecology and the Environment*, *15*(1), 42–50. doi:10.1002/fee.1451.

Iñiguez, C., Carmona, R., Lorenzo, M. R., Niell, F. X., Wiencke, C., & Gordillo, F. J. (2016). Increased temperature, rather than elevated CO 2, modulates the carbon assimilation of the Arctic kelps *Saccharina latissima* and *Laminaria solidungula*. *Marine Biology*, *163*(12), 248. doi:10.1007/s00227-016-3024-6.

Krause-Jensen, D., Lavery, P., Serrano, O., Marba, N., Masque, P., & Duarte, C. M. (2018). Sequestration of macroalgal carbon: the elephant in the Blue Carbon room. *Biology Letters*, *14*(6), 20180236. doi:10.1098/rsbl.2018.0236.

Krause-Jensen, D., & Duarte, C. M. (2016). Substantial role of macroalgae in marine carbon sequestration. *Nature Geoscience*, *9*(10), 737–742. doi:10.1038/ngeo2790.

Kuwae, T., & Hori, M. (2019). The future of blue carbon: addressing global environmental issues. *Blue Carbon in Shallow Coastal Ecosystems* (pp. 347–373). Springer, Singapore. doi:10.1007/978-981-13-1295-3_13.

Lavery, P. S., Mateo, M. Á., Serrano, O., & Rozaimi, M. (2013). Variability in the carbon storage of seagrass habitats and its implications for global estimates of blue carbon ecosystem service. *PLoS One*, *8*(9), e73748. doi:10.1371/journal.pone.0073748.

Locatelli, T., Binet, T., Kairo, J. G., King, L., Madden, S., Patenaude, G., ... & Huxham, M. (2014). Turning the tide: how blue carbon and payments for ecosystem services (PES) might help save mangrove forests. *Ambio*, *43*(8), 981–995. doi:10.1007/s13280-014-0530-y.

Lovelock, C. E., & Duarte, C. M. (2019). Dimensions of blue carbon and emerging perspectives. *Biology Letters*, *15*(3), 20180781. doi:10.1098/rsbl.2018.0781.

Macreadie, P. I., Nielsen, D. A., Kelleway, J. J., Atwood, T. B., Seymour, J. R., Petrou, K., … & Ralph, P. J. (2017). Can we manage coastal ecosystems to sequester more blue carbon?. *Frontiers in Ecology and the Environment*, *15*(4), 206–213. 206–213. doi:10.1002/fee.1484.

Macreadie, P. I., Baird, M. E., Trevathan-Tackett, S. M., Larkum, A. W. D., & Ralph, P. J. (2014). Quantifying and modelling the carbon sequestration capacity of seagrass meadows—a critical assessment. *Marine Pollution Bulletin*, *83*(2), 430–439. doi:10.1016/j.marpolbul.2013.07.038.

Macreadie, P. I., York, P. H., Sherman, C. D., Keough, M. J., Ross, D. J., Ricart, A. M., & Smith, T. M. (2014). No detectable impact of small-scale disturbances on 'blue carbon' within seagrass beds. *Marine Biology*, *161*(12), 2939–2944. doi:10.1007/s00227-014-2558-8.

Macreadie, P. I., Hughes, A. R., & Kimbro, D. L. (2013). Loss of 'blue carbon' from coastal salt marshes following habitat disturbance. *PLoS One*, *8*(7), e69244. doi:10.1371/journal.pone.0069244.

Oreska, M. P., Wilkinson, G. M., McGlathery, K. J., Bost, M., & McKee, B. A. (2018). Non-seagrass carbon contributions to seagrass sediment blue carbon. *Limnology and Oceanography*, *63*(S1), S3–S18. doi:10.1002/lno.10718.

Oreska, M. P., McGlathery, K. J., & Porter, J. H. (2017). Seagrass blue carbon spatial patterns at the meadow-scale. *PLoS One*, *12*(4), e0176630. doi:10.1371/journal.pone.0176630.

Pedersen, M. F., Filbee-Dexter, K., Norderhaug, K. M., Fredriksen, S., Frisk, N. L., & Wernberg, T. (2020). Detrital carbon production and export in high latitude kelp forests. *Oecologia*, *192*(1), 227–239. doi:10.1007/s00442-019-04573-z.

Pessarrodona, A., Moore, P. J., Sayer, M. D., & Smale, D. A. (2018). Carbon assimilation and transfer through kelp forests in the NE Atlantic is diminished under a warmer ocean climate. *Global Change Biology*, *24*(9), 4386–4398. doi:10.1111/gcb.14303.

Smale, D. A., Moore, P. J., Queirós, A. M., Higgs, S., & Burrows, M. T. (2018). Appreciating interconnectivity between habitats is key to blue carbon management. *Frontiers in Ecology and the Environment*, *16*(2), 71–73. doi:10.1002/fee.1765.

Smale, D. A., Burrows, M. T., Moore, P., O'Connor, N., & Hawkins, S. J. (2013). Threats and knowledge gaps for ecosystem services provided by kelp forests: a northeast Atlantic perspective. *Ecology and Evolution*, *3*(11), 4016–4038. doi:10.1002/ece3.774.

Sondak, C. F., Ang, P. O., Beardall, J., Bellgrove, A., Boo, S. M., Gerung, G. S., … & Largo, D. (2017). Carbon dioxide mitigation potential of seaweed aquaculture beds (SABs). *Journal of Applied Phycology*, *29*(5), 2363–2373. doi:10.1007/s10811-016-1022-1.

Sondak, C. F., & Chung, I. K. (2015). Potential blue carbon from coastal ecosystems in the Republic of Korea. *Ocean Science Journal*, *50*(1), 1–8. doi:10.1007/s12601-015-0001-9.

Thompson, B. S., Clubbe, C. P., Primavera, J. H., Curnick, D., & Koldewey, H. J. (2014). Locally assessing the economic viability of blue carbon: A case study from Panay Island, the Philippines. *Ecosystem Services*, *8*, 128–140. doi:10.1016/j.ecoser.2014.03.004.

Trevathan-Tackett, S. M., Kelleway, J., Macreadie, P. I., Beardall, J., Ralph, P., & Bellgrove, A. (2015). Comparison of marine macrophytes for their contributions to blue carbon sequestration. *Ecology*, *96*(11), 3043–3057. doi:10.1890/15-0149.1.

Vanderklift, M. A., Marcos-Martinez, R., Butler, J. R., Coleman, M., Lawrence, A., Prislan, H., … & Thomas, S. (2019). Constraints and opportunities for market-based finance for the restoration and protection of blue carbon ecosystems. *Marine Policy*, *107*, 103429. doi:10.1016/j.marpol.2019.02.001.

Wylie, L., Sutton-Grier, A. E., & Moore, A. (2016). Keys to successful blue carbon projects: lessons learned from global case studies. *Marine Policy*, *65*, 76–84. doi:10.1016/j.marpol.2015.12.020.

Zarate-Barrera, T. G., & Maldonado, J. H. (2015). Valuing blue carbon: carbon sequestration benefits provided by the marine protected areas in Colombia. *PLoS ONE*, *10*(5), e0126627. doi:10.1371/journal.pone.0126627.

Thomas's meta-analysis

References to the 46 articles in Thomas's meta-analysis.

Alongi, D. M. (2012). Carbon sequestration in mangrove forests. *Carbon Management*, *3*(3), 313–322. doi:10.4155/cmt.12.20.

Alongi, D. M. (2011). Carbon payments for mangrove conservation: ecosystem constraints and uncertainties of sequestration potential. *Environmental Science & Policy*, *14*(4), 462–470. doi:10.1016/j.envsci.2011.02.004.

Beaudoin, Y., & Pendleton, L. (2012). *Why value the oceans: a discussion paper.* UNEP/GRID-Arendal/Nicholas Institute for Environmental Policy Solutions/UNEP TEEB/UNEP Regional Seas Programme. http://www.teebweb.org/media/2013/10/2013-Why-Value-the-Oceans-Discussion-Paper.pdf

Bell, J., & Lovelock, C. E. (2013). Insuring mangrove forests for their role in mitigating coastal erosion and storm-surge: an Australian case study. *Wetlands*, *33*(2), 279–289. doi:10.1007/s13157-013-0382-4.

Benessaiah, K. (2012). Carbon and livelihoods in Post-Kyoto: assessing voluntary carbon markets. *Ecological Economics*, *77*, 1–6. doi:10.1016/j.ecolecon.2012.02.022.

Bianchi, T. S., Allison, M. A., Zhao, J., Li, X., Comeaux, R. S., Feagin, R. A., & Kulawardhana, R. W. (2013). Historical reconstruction of mangrove expansion in the Gulf of Mexico: linking climate change with carbon sequestration in coastal wetlands. *Estuarine, Coastal and Shelf Science*, *119*, 7–16. doi:10.1016/j.ecss.2012.12.007.

Broadhead, J. S. (2011). *Reality check on the potential to generate income from mangroves through carbon credit sales and payments for environmental services.* Regional Fisheries Livelihoods Programme for South and Southeast Asia. http://bluecarbonportal.org/wp-content/uploads/2012/08/Broadhead_Reality-cheack_2011.pdf

Chmura, G. L. (2013). What do we need to assess the sustainability of the tidal salt marsh carbon sink?. *Ocean & Coastal Management*, *83*, 25–31. doi:10.1016/j.ocecoaman.2011.09.006.

Crooks, S., Herr, D., Tamelander, J., Laffoley, D., & Vandever, J. (2011). *Mitigating climate change through restoration and management of coastal wetlands and near-shore marine ecosystems: challenges and opportunities.* The World Bank/IUCN/ESA PWA. https://openknowledge.worldbank.org/bitstream/handle/10986/18318/605780REPLACE-M10of0Coastal0Wetlands.pdf?sequence=1&isAllowed=y

Crooks, S., Emmett-Mattox, S., & Findsen, J. (2010). *Findings of the National Blue Ribbon Panel on the Development of a Greenhouse Gas Offset Protocol for Tidal Wetlands Restoration and Management: action plan to guide protocol development.* Restore America's Estuaries. https://estuaries.org/wp-content/uploads/2019/02/RAE-Action-Plan-to-Guide-Protocol-Development.pdf

da Silva Copertino, M. (2011). Add coastal vegetation to the climate critical list. *Nature*, *473*(7347), 255–255. doi:10.1038/473255a.

Daniel, D. B., & Willingham, D. T. (2012). Electronic textbooks: why the rush? *Science*, *335*(6076), 1569–1571. doi:10.1126/science.335.6076.1569.

Danone Fund for Nature. (2010). *Achieving carbon offsets through mangroves and other wetlands*. Danone Group/IUCN/Ramsar Convention. https://www.ramsar.org/sites/default/files/documents/pdf/DFN_report_Final.pdf

Donato, D. C., Kauffman, J. B., Murdiyarso, D., Kurnianto, S., Stidham, M., & Kanninen, M. (2011). Mangroves among the most carbon-rich forests in the tropics. *Nature Geoscience*, *4*(5), 293–297. doi:10.1038/ngeo1123.

Donato, D. C., Kauffman, J. B., Mackenzie, R. A., Ainsworth, A., & Pfleeger, A. Z. (2012). Whole-island carbon stocks in the tropical Pacific: implications for mangrove conservation and upland restoration. *Journal of Environmental Management*, *97*, 89–96. doi:10.1016/j.jenvman.2011.12.004.

Duarte, C. M., Kennedy, H., Marbà, N., & Hendriks, I. (2013). Assessing the capacity of seagrass meadows for carbon burial: current limitations and future strategies. *Ocean & Coastal Management*, *83*, 32–38. doi:10.1016/j.ocecoaman.2011.09.001.

Ellison, J. C. (2012). *Climate change vulnerability assessment and adaptation planning for mangrove systems*. World Wide Fund for Nature. https://c402277.ssl.cf1.rackcdn.com/publications/382/files/original/Climate_Change_Vulnerability_Assessment_and_Adaptation_Planning_for_Mangrove_Systems.pdf?1345748387

Fourqurean, J. W., Duarte, C. M., Kennedy, H., Marbà, N., Holmer, M., Mateo, M. A., ... & Serrano, O. (2012). Seagrass ecosystems as a globally significant carbon stock. *Nature Geoscience*, *5*(7), 505–509. doi:10.1038/ngeo1477.

Gordon, D., Murray, B. C., Pendleton, L., & Victor, B. (2011). *Financing options for blue carbon: opportunities and lessons from the REDD+ experience*. Nicholas Institute for Environmental Policy Solutions. https://nicholasinstitute.duke.edu/sites/default/files/publications/financing-options-for-blue-carbon-paper.pdf

Grimsditch, G., Alder, J., Nakamura, T., Kenchington, R., & Tamelander, J. (2013). The blue carbon special edition—introduction and overview. *Ocean & Coastal Management*, *83*, 1–4. doi:10.1016/j.ocecoaman.2012.04.020.

Hastings, J., Thomas, S., Burgener, V., Gjerde, K., Laffoley, D., Salm, R., ... & Hemley, G. (2012). Safeguarding the blue planet: six strategies for accelerating ocean protection. *Parks*, *18*(1), 9.

Herr, D., Pidgeon, E., & Laffoley, D. D. A. (2012). *Blue carbon policy framework 2.0: based on the discussion of the International Blue Carbon Policy Working Group*. IUCN. https://portals.iucn.org/library/sites/library/files/documents/2012-016.pdf

Herr, D., Pidgeon, E., & Laffoley, D. (2011). *Blue carbon policy framework: based on the first workshop of the International Blue Carbon Policy Working Group*. IUCN. https://portals.iucn.org/library/sites/library/files/documents/2011-058.pdf

Hopkinson, C. S., Cai, W. J., & Hu, X. (2012). Carbon sequestration in wetland dominated coastal systems—a global sink of rapidly diminishing magnitude. *Current Opinion in Environmental Sustainability*, *4*(2), 186–194. doi:10.1016/j.cosust.2012.03.005.

IOC-UNESCO, IMO, FAO, & UNDP. (2011). *A Blueprint for Ocean and Coastal Sustainability*. IOC/UNESCO. https://www.undp.org/content/undp/en/home/librarypage/environment-energy/water_governance/a_blueprint_for_oceanand-coastalsustainability.html

Kauffman, J. B., & Donato, D. C. (2012). *Protocols for the measurement, monitoring and reporting of structure, biomass, and carbon stocks in mangrove forests*. CIFOR. https://www.cifor.org/publications/pdf_files/WPapers/WP86CIFOR.pdf

Laffoley, D., & Grimsditch, G. D. (Eds.). (2009). *The management of natural coastal carbon sinks*. IUCN. https://www.oceanfdn.org/sites/default/files/Laffoley%20The%20Management%20of%20Natural%20Coastal%20Carbon%20Sinks-.pdf

Lau, W. W. (2013). Beyond carbon: conceptualizing payments for ecosystem services in blue forests on carbon and other marine and coastal ecosystem services. *Ocean & Coastal Management*, *83*, 5–14. doi:10.1016/j.ocecoaman.2012.03.011.

Lawrence, A. (2012). *Blue carbon: a new concept for reducing the impacts of climate change by conserving coastal ecosystems in the coral triangle.* World Wide Fund for Nature. https://wwfint.awsassets.panda.org/downloads/wwf_bluecarbon_report_coraltriangle.pdf

Lovelock, C. E., Ruess, R. W., & Feller, I. C. (2011). CO_2 efflux from cleared mangrove peat. *PLoS One*, *6*(6), e21279. doi:10.1371/journal.pone.0021279.

Luisetti, T., Jackson, E. L., & Turner, R. K. (2013). Valuing the European 'coastal blue carbon' storage benefit. *Marine Pollution Bulletin*, *71*(1–2), 101–106. doi:10.1016/j.marpolbul.2013.03.029.

Macreadie, P. I., Allen, K., Kelaher, B. P., Ralph, P. J., & Skilbeck, C. G. (2012). Paleoreconstruction of estuarine sediments reveal human-induced weakening of coastal carbon sinks. *Global Change Biology*, *18*(3), 891–901. doi:10.1111/j.1365-2486.2011.02582.x.

Matsui, N., Morimune, K., Meepol, W., & Chukwamdee, J. (2012). Ten year evaluation of carbon stock in mangrove plantation reforested from an abandoned shrimp pond. *Forests*, *3*(2), 431–444. doi:10.3390/f3020431.

Mcleod, E., Chmura, G. L., Bouillon, S., Salm, R., Björk, M., Duarte, C. M., … & Silliman, B. R. (2011). A blueprint for blue carbon: toward an improved understanding of the role of vegetated coastal habitats in sequestering CO_2. *Frontiers in Ecology and the Environment*, *9*(10), 552–560. doi:10.1890/110004.

Murray, B. C., Pendleton, L., Jenkins, W. A., & Sifleet, S. (2011). *Green payments for blue carbon: economic incentives for protecting threatened coastal habitats.* Nicholas Institute for Environmental Policy Solutions. https://nicholasinstitute.duke.edu/sites/default/files/publications/blue-carbon-report-paper.pdf

National Oceanic and Atmospheric Association. (2011). *Opportunities to use carbon services to advance coastal habitat conservation.* http://www.bluecarbonportal.org/wp-content/uploads/2012/08/NOAA_Opportunities-to-Advance-Coastal-Habitat-Conservation_June2011.pdf

Nelleman, C., Corcoran, E., Duarte, C. M., Valdes, L., DeYoung, C., Fonseca, L., & Grimsditch, G. (2008). *Blue carbon: the role of healthy oceans in binding carbon.* UNEP/FAO/UNESCO/IUCN/CSIC. https://ccom.unh.edu/sites/default/files/publications/Nellemann_2010_BlueCarbon_book.pdf

O'Sullivan, R., Chagas, T., Burns, D., Pearson, T. (2011). *Blue carbon policy options assessment*, 4th ed. Climate Focus. https://climatefocus.com/sites/default/files/blue_carbon_.pdf

Pendleton, L., Donato, D. C., Murray, B. C., Crooks, S., Jenkins, W. A., Sifleet, S., … & Megonigal, P. (2012). Estimating global "blue carbon" emissions from conversion and degradation of vegetated coastal ecosystems. *PLoS One*, 7(9), e43542. doi:10.1371/journal.pone.0043542.

Russell, B. D., Connell, S. D., Uthicke, S., Muehllehner, N., Fabricius, K. E., & Hall-Spencer, J. M. (2013). Future seagrass beds: can increased productivity lead to increased carbon storage?. *Marine Pollution Bulletin*, *73*(2), 463–469. doi:10.1016/j.marpolbul.2013.01.031.

Sidik, F., & Lovelock, C. E. (2013). CO_2 efflux from shrimp ponds in Indonesia. *PLoS One*, *8*(6), e66329. doi:10.1371/journal.pone.0066329.

Sifleet, S., Pendleton, L., & Murray, B. C. (2011). *State of the science on coastal blue carbon: a summary for policy makers.* Nicholas Institute for Environmental Policy Solutions. https://nicholasinstitute.duke.edu/sites/default/files/publications/state-of-science-coastal-blue-carbon-paper.pdf

Siikamäki, J., Sanchirico, J. N., & Jardine, S. L. (2012). Global economic potential for reducing carbon dioxide emissions from mangrove loss. *Proceedings of the National Academy of Sciences, 109*(36), 14369–14374. doi:10.1073/pnas.1200519109.

Sutton-Grier, A. E., Moore, A. K., Wiley, P. C., & Edwards, P. E. (2014). Incorporating ecosystem services into the implementation of existing US natural resource management regulations: operationalizing carbon sequestration and storage. *Marine Policy, 43*, 246–253. doi:10.1016/j.marpol.2013.06.003.

Ullman, R., Bilbao-Bastida, V., & Grimsditch, G. (2013). Including blue carbon in climate market mechanisms. *Ocean & Coastal Management, 83*, 15–18. doi:10.1016/j.ocecoaman.2012.02.009.

Yee, S. (2010). REDD and BLUE carbon: carbon payments for mangrove conservation. *Master's thesis*, Center for Marine Biodiversity and Conservation, University of California San Diego. https://escholarship.org/uc/item/2bc6j8p

15

STRENGTHENING INDUSTRY AND ACADEMIC LINKS THROUGH TRANSDISCIPLINARY ACTION RESEARCH

An Introspective Reflection of a Collaborative Water Quality, Biodiversity, and Aquaculture Initiative

Lauren Brideau, Emma L. Cross, Annette F. Govindarajan, Dan Martino, Greg Martino, Miranda Holland, Gabriela Triay, and C. Patrick Heidkamp

List of Abbreviations

- Transdisciplinary action research (TAR)
- Integrated multi-trophic aquaculture (IMTA)

15.1 Introduction

15.1.1 Multidimensional Solutions to Coastal Challenges

Coastal regions are complex, dynamic, and support socio-ecological communities that provide fundamental ecosystem services. However, anthropogenic stressors lead to adverse impacts on coastal and marine systems (Bennett et al., 2019; Custódio et al., 2020) placing these zones at the forefront of the struggle for sustainability (Heidkamp & Morrissey, 2018). The Blue Economy has emerged as a dynamic pathway to facilitate growth of social and economic sectors while sustaining ocean resilience, particularly as answers to economic development focus on ocean-related solutions (Bennett et al., 2019; McKinley, 2023 (this volume); Garland et al., 2019). Furthermore, there is a growing consensus that engaging actors impacted by Blue Economy initiatives is required for just and sustainable transitions (Campbell et al., 2021; Heidkamp et al., 2021; McKinley, 2023 (this volume)). This finding suggests that inclusion of scientific and non-scientific sectors, a concept and methodology often referred to as transdisciplinary action research (TAR), is beneficial in driving the Blue Economy agenda forward.

DOI: 10.4324/9781003280248-19

Aquaculture is recognised by the European Commission (2017) as one of five sectors holding a promising pathway to advance the Blue Economy through sustainable jobs and blue growth. Furthermore, with aquaculture leading expansion of global food production coupled with predictions of growing demands for seafood (Campbell et al., 2021), marine aquaculture in seawater and brackish water is increasingly recognised as an agent to expand Blue Economy initiatives (Ahmed & Thompson, 2019). To accomplish such tasks and mitigate environmental degradation along with social injustices, proper aquaculture management and design is required (Troell et al., 2014). While some aquaculture practices may cause adverse impacts, certain practices can increase equity (Troell et al., 2014) and ecosystem services (Custódio et al., 2020) while fulfilling demands for seafood. For example, integrated multi-trophic aquaculture (IMTA) grows fed species (such as finfish or shrimp) and extractive species (such as seaweed or shellfish) with the aim of balancing by-products of fed species as feed for extractive species. IMTA works toward the concept that properly functioning co-cultured aquaculture systems are more sustainable than monocultures (Ahmed & Thompson, 2019; Barrington et al., 2009; Custódio et al., 2020). Technologies in marine aquaculture such as IMTA have emerged as methods to increase ocean resilience by diversifying species (Ahmed & Thompson, 2019; Troell et al., 2014), thus working toward ecosystem approaches for social, economic, and ecological benefits.

Multi-species ocean farming uses the water column to co-culture shellfish and seaweed. Shellfish filter organic waste and nutrients from the water which may reduce eutrophication (Ahmed & Thompson, 2019; Troell et al., 2014) and increase the ability of seaweeds to photosynthesise. Seaweeds, in turn, have the potential to sequester carbon while raising oxygen and pH levels during photosynthesis which could benefit shellfish growth (Duarte et al., 2017). Consequently, multi-species ocean farming offers a mechanism for mutually beneficial relationships between co-cultured species, and unlike IMTA, does not require feed or fertiliser making it a more sustainable form of protein farming. Aquaculture practices like multi-species ocean farming may prevent ocean degradation that results from viewing the Blue Economy exclusively as an economic resource (McKinley, 2023) and supports coastal communities (Campbell et al., 2021) as sustainability efforts are enhanced by considering human dimensions (Bennett, 2019). To examine social dimensions of coastal and marine challenges (Bennett, 2019), drawing on marine social sciences becomes paramount to uniting different subjects, research pathways, and disciplines (McKinley, 2023). Such a method involves integrating all stakeholders in processes and considering the mechanisms, benefits, and end goals that each stakeholder is interested in.

TAR has been set forth as an emerging and evolving method to solve real-world problems through diverse partnerships (Lang et al., 2012; Steger et al., 2021) by involving "science with society" (Heidkamp et al., 2021; Steger et al.,

2021). By not restricting stakeholder engagement (Heidkamp et al., 2021) and building on equitable partnerships to co-learn and co-produce knowledge, TAR moves beyond traditional collaborations, which are often dominated by uneven relations. While TAR approaches have tremendous potential to resolve these challenges, they have been labelled as inherently "messy" (Heidkamp et al., 2021) and limited frameworks may not encompass the complexity of many project-based practices. Lang et al. (2012) set forth a TAR framework of three phases: (a) problem framing and team building, (b) co-creation of knowledge, and (c) re-integration and application of knowledge. Heidkamp et al. (2021) argue for integration of government, academia, business, society, and inclusive TAR for stakeholder engagement, while Steger et al. (2021) present seven steps aiming to generate co-production and co-learning. Importantly, regardless of proposed frameworks, the inclusion of diverse actors underpins the foundation of TAR approaches. These methods have been applied to various case studies including land-use change (Huber & Rigling, 2014), socio-technical transformation (Heidkamp et al., 2021), and solid waste management (Chammas et al., 2020). While accessing a TAR cycle can be accomplished by engaging with applied topics (Heidkamp et al., 2021), few studies report on evidence of best practices (Huber & Rigling, 2014) and overcoming challenges in TAR is ambiguous (Steelman et al., 2015). Multi-species ocean farming is an emerging research agenda in the Blue Economy narrative and its sustainable practices for increasing ocean health, economic valuation, and services that benefit and involve multiple stakeholders create a platform where Blue Economy agendas can be recognised through TAR. Here, we aim to co-produce TAR knowledge on challenges and best practices through a reflection on an applied TAR multi-species ocean farming project that we initiated. This reflection consists of a synthesis created by the lead author (L. Brideau) of collaborator (co-author) self-reflections on our TAR project in the context of a generalised TAR model (Heidkamp et al., 2021) as described below.

15.2 Our TAR Project

Our TAR study was started by an idea generated at the 2020 National Seaweed Symposium in Providence, Rhode Island between industry partners at Cottage City Oysters (co-authors D. and G. Martino) and academic partners at Southern Connecticut State University (SCSU; co-authors Brideau, Cross, Holland, Triay, and Heidkamp). After initial meetings between Cottage City Oysters and the SCSU team, additional expertise, from an academic researcher at the Woods Hole Oceanographic Institution (WHOI; co-author Govindarajan), subsequently (after the Symposium) joined the project.

An early aim of our TAR project was to explore the feasibility for an eco-label that highlights ecological benefits (biodiversity, carbon sequestration potential, etc.) of multi-species ocean farming. Thus, a shared interest in multi-species ocean farming developed into an applied water quality, biodiversity, and aquaculture research initiative to concurrently evaluate sustainable social, economic, and environmental practices. We selected continuous water quality measurements and environmental DNA (eDNA) analysis to study biological and chemical parameters at Cottage City Oysters' new farm site before and after implementing oyster cages and kelp lines. Located off the coast of Martha's Vineyard, the United States, a previously unfarmed ocean plot offered a unique opportunity to assess ecosystem changes associated with an emerging technique for sustainable seafood production. It is important to understand how multi-species ocean farming impacts the marine environment, in terms of potential enhanced ecosystem services and reduction of adverse impacts, particularly as coastal and marine systems become increasingly vulnerable to climate impacts.

Despite the importance of strengthening the market value of habitat provisioning from sustainable aquaculture industry activities, applications of market values are limited (Theuerkauf et al., 2021). Thus, we explored how potential environmental benefits from multi-species ocean farming can be advanced while assessing the feasibility of a biodiversity ecolabel, a mark to identify products complying with environmental production standards (US EPA, 2014). An ecolabel could create a market niche for sustainable aquaculture products grown in multi-species ocean farms while ensuring farmers receive a price premium for adhering to production standards. Furthermore, ecolabels allow information about positive environmental benefits of multi-species ocean farming to be transmitted to the consumer. Our applied TAR project balanced goals of business and academic sectors by working to advance Blue Economy initiatives through acquired knowledge (Table 15.1).

15.3 Our Process

To concurrently assess TAR from academic and industry perspectives, this chapter was collaboratively written as an introspective reflection with the goal of co-producing insight on our applied TAR project. The TAR process was evaluated by the conceptual model of action research cycles (Figure 15.1) established by Heidkamp et al. (2021) and supported by guiding questions for transdisciplinary research project design principles outlined by Lang et al. (2012). Since our study is in its early stages (year 1.5 out of an anticipated minimum of 3), a subset of actions in the Heidkamp et al. (2021) model was analysed:

- Phase 1: Engage with Real-World Context and Define Purpose (aligns with Phase A: problem framing and team building from Lang et al., 2012)
- Phase 2: Plan Action, Take Action, Evaluate Action (aligns with Phase B: co-creation of solution-oriented transferable knowledge from Lang et al., 2012)
- Unexpected Events
- Desired Outcome

TABLE 15.1 Framework of project motivations and anticipated timeline for objectives

Project aim	*Objective*	*Steps*	*Outcome*	*Timeline (years)*
Quantify chemical and biological impacts of multi-species ocean farming, and generate a biodiversity ecolabel	Equitable involvement and collaboration among academic and societal actors	TAR	Partnerships between industry, academia, students, and local collaborators	1–3 and beyond
	Quantify biological and chemical changes of farm site	Collect and analyse data	Inform aquaculture management and practices	1–3
	Assess feasibility of an ecolabel	Evaluate environmental impacts	Generate ecolabel if positive environmental impacts are quantified	3
	Dissemination to local, scientific, and aquaculture industry communities	Communicate results	Presentations, publications, podcasts	1–3

15.3.1 Phase 1: Engage with Real-World Context and Define Purpose

The first step in our TAR project's initiation was integrating applied and scientific goals, relevant to team members' interests. While TAR projects may stem from researchers, practitioners, or stakeholders (Steger et al., 2021), our project was concurrently initiated by researchers and practitioners. Collaboration and project scheme were generated at a national seaweed symposium where we (industry and academic personnel) discussed multi-species ocean farming in a breakout group:

> The lack of an (multi-species ocean farm biodiversity) ecolabel for the aquaculture industry struck us both as odd and an inevitable endeavour that someone would create. We agreed how important it would be that such a label be science and data driven, and not an arbitrary label handed out by a corporation to serve their marketing purpose. We decided to take on the responsibility for the creation of the first industry ecolabel because then, and only then, could we guarantee that the process would be scientifically sound, data driven, and based on a real-world model.—P1

The initial seaweed symposium meeting led to partial composition of the research team with further team members, crucial to the project's success, joining after

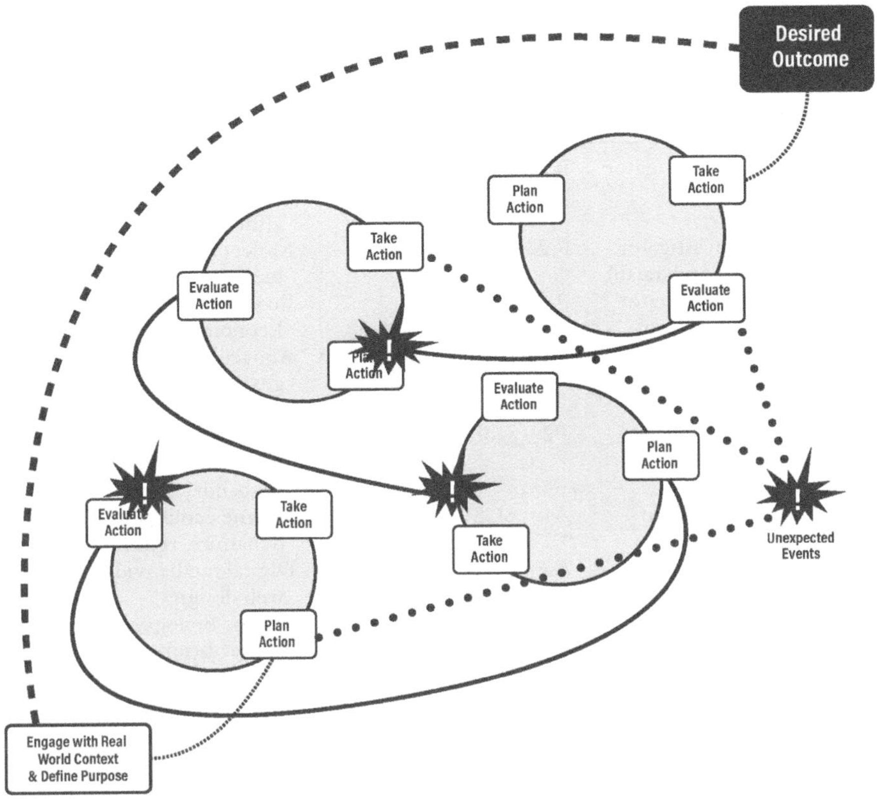

FIGURE 15.1 Reality of TAR cycle encompassing challenges and unexpected events (Heidkamp et al., 2021)

the symposium. Since TAR projects require specific expertise in numerous disciplines, team building following an initial TAR proposal is often recursive (Lang et al., 2012). Our full transdisciplinary team includes industry partners, an interdisciplinary cohort of academic researchers, research assistants, and students (Table 15.2). In our reflection, we refer to ourselves by the codes listed in this table.

Our team members are located across two states and three cities. We represent varying professional stages and disciplines allowing a range of integrated geographies, methodologies, and perspectives to inform aquaculture in the Blue Economy. All of us have previous experience in environmental work or research which generated commitment to the project. Our previous knowledge of the study system also has the advantage of addressing current challenges and issues that may be more difficult for people lacking previous involvement in a study to obtain, such as equipment needs, current knowledge of problems and solutions, and an established network surrounding a project's focus.

Our TAR project had three broad motivations: understanding ecological transitions, advancing economic valuation, and developing collaborative TAR

TABLE 15.2 Research team members, affiliation, and area of expertise related to the project. Code denotes reference to collaborator used throughout the chapter. Table format adapted from Gaziulusoy et al. (2016)

Collaborator	*Code*	*Base*	*Project expertise*
Lead principal investigator (co-author Cross)	R1	SCSU	Marine ecology, environmental monitoring, climate change
Co-principal investigator (co-author Govindarajan)	R2	WHOI	Molecular ecology, eDNA, biodiversity
Co-principal investigator (co-author Heidkamp)	R3	SCSU	Economic geography, Blue Economy, consumer insight
Industry professional 1 (co-author D. Martino)	P1	Cottage city oysters	Aquaculture industry, farm gear management, product husbandry
Industry professional 2 (co-author G. Martino)	P2	Cottage city oysters	Aquaculture industry, farm gear management, product husbandry
Research assistant (lead author L. Brideau)	A1	SCSU	Marine ecology, coastal dynamics, restoration
Graduate assistant (co-author G. Triay)	A2	SCSU	Digital media, videography, web design
Undergraduate assistant (co-author M. Holland)	A3	SCSU	Marine biology, biodiversity, conservation

approaches, and our interest in the study varied across these themes. As a marine ecologist, R1 was focused on environmental monitoring. R2 was particularly interested in understanding how biodiversity changes as a result of human activities while R3 was motivated by facilitating partnerships and economic applications of research. R3 was also motivated by the possibilities a TAR approach might provide in terms of research opportunities for both undergraduate and graduate students.

> I believe that TAR provides a tremendous opportunity to link academic research to practical applications, but it also provides a framework that allows for the integration of student researchers from differing disciplines and at different levels of their research career.—R3

Interests of research assistants and students closely aligned with those of team researcher leads. Industry collaborators P1 and P2 were primarily driven to understand the quantity of biodiversity attracted to aquaculture farm sites with various types of farm gear and the baseline chemistry of unfarmed ocean plots. P1 and P2 were also interested in how results of this data could generate an ecolabel and therefore higher product prices. While specific interests within the study varied, our research questions were able to be explored across different sectors and disciplines.

Our varied but intersecting academic and industry perspectives aligned with our shared research goal to advance knowledge surrounding environmental impacts of aquaculture. While we had a shared understanding and interest in the principal research problem, we simultaneously set forth a multifaceted research purpose allowing different sectors to prioritise discrete goals. Thus, a commitment to the overarching research problem was upheld while each collaborator individually approached the problem from their area of expertise allowing for a broader range of topics to be integrated in the project. For example, our study targeted socio-ecological challenges with an economic implementation interest that fell within multiple sustainable development goals: 12—responsible consumption and production, 13—climate action, 14—life below water, and 17—partnerships for the goals.

15.3.2 Phase 2: Plan Action, Take Action, and Evaluate Action

We used a solution-oriented approach to apply TAR to how practitioners may enhance and create more resilient water quality and biodiversity parameters through multi-species ocean farming. After establishing research goals and methods in Phase 1, we agreed on collecting biological and chemical data to advance our project's aim. This was a critical step underpinning our project's objectives since understanding how surrounding marine communities respond to multi-species ocean farming prior to knowledge implementation is necessary. Furthermore, this approach broadened the exploration of applied social and scientific research questions and linked our project to implementation interests.

We developed a broad methodological framework to investigate the research problem allowing project design evolution in response to challenges. Team members took responsibility for achieving tasks in their research or business domain by furthering the methodological framework, preparing equipment, and creating protocols to move the research design forward. P1 and P2 conducted all preliminary field activities prior to scientific monitoring or sampling. This included securing the ocean plot to be studied and deploying buoys for the water quality probes. R2 developed an eDNA protocol and built a custom sampler allowing field researchers to take water samples while reducing the potential for contamination—a necessity when working with eDNA. R1 and R3 secured multiprobes to monitor water quality, gathered necessary water quality research equipment, and facilitated the first fieldwork campaign.

While conducting research, each member maintained a specific role aligned with their area of expertise and thus interest in the project. P1 and P2 operated the research site and facilitated collection of eDNA and calibration of the monitoring probes. Furthermore, P1 and P2 deployed and maintained all aquaculture gear and organisms associated with the site. R2 led eDNA research by processing eDNA samples collected monthly from two sampling locations with the goal of assessing how the community changes over time as a result of aquaculture activities. R1 coordinated monthly fieldwork campaigns, led field and lab water quality analysis, and led the writing of grant proposals to obtain funding. R3 directed research implementation and fostered industry and academic partnerships. Despite established project roles, in application, these roles were partially fluid as we assisted each other in various ways on all aspects of the project, which facilitated co-learning and co-creation of knowledge. However, early research stages revealed that some tasks were sector specific and non-transferable to collaborators. We found that transferable and non-transferable tasks exerted strong controls on our individual involvement in reaching the project's objectives. Non-transferable tasks primarily emerged as a function of time constraints, geographic locations, equipment needs, and required training in advanced and specific knowledge to complete tasks. Examples of these tasks include filtering seawater for eDNA samples and maintaining aquaculture sites.

Evaluation is a critical step in TAR, and we found it can result in rescaling a project by increasing a project's detail. A primary outcome that emerged from our TAR objective to increase involvement was the addition of two short-term undergraduate student projects. In evaluating actions and associated timelines, we were able to assess the feasibility of implementing student projects which generated a more detailed study. While student projects had a shorter timeline (semester to year) and were a subset of the overall project, students acquired real-world experience on questions related to their interests and field of study.

> I became part of the project after asking R3 about ideas for my honors thesis. R3 suggested I set up a meeting with R1 to discuss adding a shorter-term biodiversity study onto their existing project at Martha's Vineyard. My interests lie in the biodiversity analysis.—A3

15.3.3 Unexpected Events

Members in TAR often face a variety of unexpected events and challenges including inherent, institutional, practical, operational, and logistical (Gaziulusoy et al., 2016). In this study, our early-stage challenges fell into three macro-level categories that have been described in the literature: (1) teamwork challenges originating from collaborations between different sectors and subjects, (2) institutional challenges characteristic to academic institutions, and (3) emergent challenges lying beyond the control of the research team (Gaziulusoy et al., 2016). Further, we suggest the concept "societal challenges" to encompass non-academic challenges that are inherent to TAR projects and faced by stakeholders and industry practitioners. In general, societal challenges may originate from time commitments, use of personal resources, and a lack of formal acknowledgement in the project's output. In our project, an example of a societal challenge was the flexibility required to reschedule fieldwork days due to inclement weather.

We found that challenges generate, interact with, and compound additional challenges. In our study, the COVID-19 pandemic was a principal disruption to the project that lay outside of our control. Thus, the pandemic can primarily be classified as an emergent challenge. In-person interactions were limited, and our team development and bonding were mainly online leading to less efficient communication. While previous work (Gaziulusoy et al., 2016) also found teamwork challenges early in TAR, the pandemic was unique as it generated several additional long-term challenges. For example, R1 and R2 had to manage limited lab access and attain approval for fieldwork which complicated scheduling. These issues were in addition to institutional challenges commonly outlined in TAR such as a lack of funding, timeline pressures, and difficulties for early career researchers (Gaziulusoy et al., 2016; Heidkamp et al., 2021; Polk, 2014).

Although unexpected events and challenges are characteristic of TAR, we found that they presented pathways for advantageous project adaptation by increasing flexibility to reach desired outcomes. For example, in our project, we could co-address some challenges (e.g., the pandemic), which ultimately strengthened the project by making it more resilient and flexible to stressors. Since the pandemic limited lab access, R1 and R2's teams could not interact directly. A process was developed where R2 organised sampling equipment for R1's team members to collect eDNA samples. The eDNA samples were subsequently transferred to R2 for R2's team members to process. Despite these disruptions, data collection was upheld, and fieldwork personnel abilities were expanded. While beneficial adaptations from challenges can increase a project's adaptive capacity, we found that other challenges can also present obstacles that slow down objectives or goals. For example, weather was a primary challenge that slowed our work. Although weather-related challenges are inherent in fieldwork, they are recurring; for example, on numerous occasions, weather conditions caused last-minute scheduling of our fieldwork campaigns.

15.3.4 Desired Outcomes

In our TAR project, we placed a priority on sustainability through economic, social, and environmental systems. We accomplished this by targeting sustainability-related outcomes such as community outreach, a biodiversity ecolabel and certification for multi-species aquaculture farms, and quantitative ecosystem data.

> I anticipate data that backs our hypothesis that 3d regenerative (multi-species) aquaculture farms increase biodiversity and improve ocean chemistry against climate change. I hope we can use this data to create a non-profit, ecolabel certifying organization that would certify aquaculture farms around the world for the positive ecosystem services they provide.—P1
>
> This project will provide the opportunity to assess the mitigation potential of co-culturing seaweed and shellfish for shellfish aquaculture products. In addition, this work will assess the feasibility of developing an ecolabel for sustainable aquaculture products from regenerative ocean farms (multi-species ocean farms).—R1

Prioritised identifications for each member corresponded to their work identification, research sector, and career stage. R1 and R2 identified peer-reviewed publications and presentations as anticipated project outcomes, students discussed how the project would advance their scholarship and career path, and industry professionals identified the desire to expand an ecolabel into a multitiered certification. Anticipated desired outcomes from our academic and industry members varied between disciplines, career stages, and backgrounds and translated into subcategories extending from the shared goal. Yet, all members were driven by co-producing knowledge to better the environment. As solution-oriented problem-solving serves as an overarching agenda, in our experience, different sectors can approach a single problem with varied interests and desired outcomes creating multifaceted research.

15.4 Implications

Our TAR project aligns with the Blue Economy agenda and our self-reflection has yielded insights on how TAR methods could effectively contribute to successful Blue Economy transitions. We summarise these insights in the context of the TAR pathway. First, project familiarity among collaborators or previous involvement with the research platform and shared goal can sustain long-term motivation. In our study, we were involved with the research platform and shared goal of coastal marine systems prior to the project's implementation. Our past involvement motivates our commitment to the project's future, an observation consistent with the TAR process. Prior project familiarity may relate to the importance of establishing shared goals early on in TAR, which was outlined by Steger et al. (2021) and noted as an important factor for successful TAR projects (Huber & Rigling, 2014). In our experience, given the effort and flexibility required to deal with recurring challenges (e.g., weather), we hypothesise that recurring challenges could potentially exhaust members' commitment if long-term investment in the project is not established.

Second, we found that an external shared goal falling outside the confines of collaborators' institutions or careers and relating to members' individual interests within the project can promote collaborative support across sectors. For example, we were enthusiastic about coastal marine habitats and interested in bettering the environment by advancing aquaculture technology. Consequently, our industry partners and academic researchers are providing support to each other's goals while working to advance project outputs specific to their areas of study and work. Thus, we suggest that TAR should strive for overlapping and shared responsibilities as project roles that are fluid across members can facilitate co-learning and co-creation of knowledge.

Third, we found that generating knowledge with a high implementation interest to all collaborators promotes objectives across different scales. Our project is generating knowledge with a high implementation interest to industry and academic collaborators; however, these goals also relate to social and scientific applications. For social applications, if we find that multi-species ocean farming does not harm or beneficially affects local biodiversity, we could generate an eco-label to certify our industry partners' aquaculture products. This process works to link knowledge produced by the project to implementation interests, which has been found to alleviate unbalanced problem ownership, or dominance of specific actors in projects (Lang et al., 2012). Equally, implementation interests were closely related to relevant scientific applications as multi-species ocean farming is an emerging aquaculture technique with potential ecosystem benefits. Our observations imply that implementation interests can inform knowledge on aquaculture practices for ecosystem services.

TAR offers a promising pathway to successfully transition towards sustainable Blue Economy agendas. As coastal and marine environments, and the services they provide, become increasingly threatened by anthropogenic stressors, equitable transitions toward sustainable development are required. In particular, water quality and biodiversity are two important drivers of ecosystem functioning that can help sustain social and environmental health and stability. Previous studies on Blue Economy aquaculture initiatives have focused on understanding aquaculture practices as socio-ecological systems (Johnson et al., 2019) and shifting towards "blue communities" or community welfare (Campbell et al., 2021). These studies highlight the importance of social dimensions in Blue Economies. Our transdisciplinary collaboration builds on this foundation while incorporating environmental research through a TAR framework.

15.5 Conclusion

In this chapter, we reflected upon our ongoing applied water quality, biodiversity, and aquaculture TAR project in the context of the conceptual model of TAR by Heidkamp et al. (2021) and design principles outlined by Lang et al. (2012). Our introspective reflection suggests that researching implementation interests relevant to novel or applied social and scientific endeavours, supporting shared goals, and project familiarity among collaborators can facilitate TAR progress. Our project, like every TAR project, has unique actors, goals, timelines, and unexpected events. Therefore, we suggest that adaptability should be embraced as a vehicle to project success in TAR, an insight also outlined by Heidkamp et al. (2021) and Lang et al. (2012). TAR is known as a messy process (Heidkamp et al., 2021; Steelman et al., 2015) with a variety of challenges (Gaziulusoy et al., 2016). Based on our experience and TAR project, we argue that multifaceted projects are beneficial to TAR as they create dynamic pathways for sectors to advance individual outcomes while supporting a collective goal. We also present the term "societal challenges" to encompass challenges faced by collaborators outside research institutions and suggest that this concept should be further explored in future studies on the TAR process. Our TAR team's collaborative writing process of this chapter is in itself a unique TAR component that promotes our team cohesiveness as well as contributes to the growing discourse surrounding TAR methods to advance an evolving Blue Economy agenda.

Acknowledgements

This work was supported through startup funds from the Dean of Arts & Sciences and a Southern Connecticut State University Faculty Creative Activity Research Grant awarded to E. Cross which provided funding for fieldwork travel and consumables for the water quality analysis. The Borrego Foundation and the Naktenis Foundation provided funding to A. Govindarajan for environmental DNA work. Furthermore, we thank Erin Frates, Fred Marin, Elizabeth Allan, Sara Gerckens, Maeve Rourke, and Ian Bergemann for their assistance with the project.

References

Ahmed, N., & Thompson, S. (2019). The blue dimensions of aquaculture: A global synthesis. *Science of the Total Environment, 652*, 851–861. https://doi.org/10.1016/j.scitotenv.2018.10.163

Barrington, K., Chopin, T., & Robinson, S. (2009). Integrated multi-trophic aquaculture (IMTA) in marine temperate waters. *Integrated Mariculture: A Global Review*. FAO Fisheries and Aquaculture Technical Paper, 529, 7–46.

Bennett, N. J. (2019). Marine social science for the peopled seas. *Coastal Management, 47*(2), 244–252. https://doi.org/10.1080/08920753.2019.1564958

Bennett, N. J., Cisneros-Montemayor, A. M., Blythe, J., Silver, J. J., Singh, G., Andrews, N., Calò, A., Christie, P., Di Franco, A., Finkbeiner, E. M., Gelcich, S., Guidetti, P., Harper, S., Hotte, N., Kittinger, J. N., Le Billon, P., Lister, J., López de la Lama, R., McKinley, E., … Sumaila, U. R. (2019). Towards a sustainable and equitable Blue Economy. *Nature Sustainability, 2*(11), 991–993. https://doi.org/10.1038/s41893-019-0404-1

Campbell, L. M., Fairbanks, L., Murray, G., Stoll, J. S., D'Anna, L., & Bingham, J. (2021). From Blue Economy to Blue Communities: Reorienting aquaculture expansion for community wellbeing. *Marine Policy, 124*. https://doi.org/10.1016/j.marpol.2020.104361

Chammas, G., Kayed, S., Al Shami, A., Kays, W., Citton, M., Kalot, M., Al Marj, E., Fakhr, M., Yehya, N. A., Talhouk, S. N., Al-Hindi, M., Zein-El-Dine, S., Tamim, H., Lakkis, I., Abou Najm, M., & Saliba, N. A. (2020). Transdisciplinary interventions for environmental sustainability. *Waste Management, 107*, 159–171. https://doi.org/10.1016/j.wasman.2020.03.043

Custódio, M., Villasante, S., Calado, R., & Lillebø, A. I. (2020). Valuation of ecosystem services to promote sustainable aquaculture practices. *Reviews in Aquaculture, 12*(1), 392–405. https://doi.org/10.1111/raq.12324

Duarte, C. M., Wu, J., Xiao, X., Bruhn, A., & Krause-Jensen, D. (2017). Can seaweed farming play a role in climate change mitigation and adaptation? *Frontiers in Marine Science, 4*. https://doi.org/10.3389/fmars.2017.00100

European Commission. (2017). *Towards More Sustainable Growth and Jobs in the Blue Economy (Commission Staff Working Document 128 Final)*. European Commission.

Garland, M., Axon, S., Graziano, M., Morrissey, J., & Heidkamp, C. P. (2019). The Blue Economy: Identifying geographic concepts and sensitivities. *Geography Compass*, 13(7), e12445. https://doi.org/10.1111/gec3.12445

Gaziulusoy, I., Ryan, C., McGrail, S., Chandler, P., & Twomey, P. (2016). Identifying and addressing challenges faced by transdisciplinary research teams in climate change research. *Journal of Cleaner Production, 123*, 55–64. https://doi.org/10.1016/j.jclepro.2015.08.049

Heidkamp, P., Garland, M., & Krak, L. (2021). Enacting a just and sustainable Blue Economy through transdisciplinary action research. *The Geographical Journal*. https://doi.org/10.1111/geoj.12410

Heidkamp, P., & Morrissey, J. (2018). *Towards Coastal Resilience and Sustainability* (1st ed.). Routledge.

Huber, R., & Rigling, A. (2014). Commitment to continuous research is a key factor in transdisciplinarity. Experiences from the *Mountland* project. *GAIA—Ecological Perspectives for Science and Society, 23*(3), 256–262. https://doi.org/10.14512/gaia.23.3.10

Johnson, T., Beard, K., Brady, D., Byron, C., Cleaver, C., Duffy, K., Keeney, N., Kimble, M., Miller, M., Moeykens, S., Teisl, M., van Walsum, G., & Yuan, J. (2019). A social-ecological system framework for marine aquaculture research. *Sustainability, 11*(9), 2522. https://doi.org/10.3390/su11092522

Lang, D. J., Wiek, A., Bergmann, M., Stauffacher, M., Martens, P., Moll, P., Swilling, M., & Thomas, C. J. (2012). Transdisciplinary research in sustainability science: Practice, principles, and challenges. *Sustainability Science*, 7(S1), 25–43. https://doi.org/10.1007/s11625-011-0149-x

McKinley, E. (2023). A Blue Economy for who? Linking marine social sciences with Blue Economy discourse. In C.P Heidkamp, J. Morrissey, C. Germond-Duret (Eds.), *Blue Economy: People and Regions in Transitions*. Oxford, UK: Routledge.

Polk, M. (2014). Achieving the promise of transdisciplinarity: A critical exploration of the relationship between transdisciplinary research and societal problem solving. *Sustainability Science, 9*(4), 439–451. https://doi.org/10.1007/s11625-014-0247-7

Steelman, T., Nichols, E. G., James, A., Bradford, L., Ebersöhn, L., Scherman, V., Omidire, F., Bunn, D. N., Twine, W., & McHale, M. R. (2015). Practicing the science of sustainability: The challenges of transdisciplinarity in a developing world context. *Sustainability Science, 10*(4), 581–599. https://doi.org/10.1007/s11625-015-0334-4

Steger, C., Klein, J. A., Reid, R. S., Lavorel, S., Tucker, C., Hopping, K. A., Marchant, R., Teel, T., Cuni-Sanchez, A., Dorji, T., Greenwood, G., Huber, R., Kassam, K.-A., Kreuer, D., Nolin, A., Russell, A., Sharp, J. L., Šmid Hribar, M., Thorn, J. P. R., … Waiswa, D. (2021). Science with society: Evidence-based guidance for best practices in environmental transdisciplinary work. *Global Environmental Change, 68*(10), 102240. https://doi.org/10.1016/j.gloenvcha.2021.102240

Theuerkauf, S. J., Barrett, L. T., Alleway, H. K., Costa-Pierce, B. A., St. Gelais, A., & Jones, R. C. (2021). Habitat value of bivalve shellfish and seaweed aquaculture for fish and invertebrates: Pathways, synthesis and next steps. *Reviews in Aquaculture*, raq.12584. https://doi.org/10.1111/raq.12584

Troell, M., Naylor, R. L., Metian, M., Beveridge, M., Tyedmers, P. H., Folke, C., Arrow, K. J., Barrett, S., Crépin, A.-S., Ehrlich, P. R., Gren, Å., Kautsky, N., Levin, S. A., Nyborg, K., Österblom, H., Polasky, S., Scheffer, M., Walker, B. H., Xepapadeas, T., & de Zeeuw, A. (2014). Does aquaculture add resilience to the global food system? *Proceedings of the National Academy of Sciences, 111*(37), 13257–13263. https://doi.org/10.1073/pnas.1404067111

US EPA, O. (2014). *Introduction to Ecolabels and Standards for Greener Products [Overviews and Factsheets]*. https://www.epa.gov/greenerproducts/introduction-ecolabels-and-standards-greener-products

16

CONCLUSION

Situating Just Transitions: Sustainability, Innovation, and Inclusion in the Blue Economy?

C. Patrick Heidkamp, Michaela Garland, John Morrissey, Celine Germond-Duret, and Matthias Kokorsch

16.1 Premise

It is fitting that the writings collected in this edited volume coincide with the start of the UN's Decade of Ocean Science for Sustainable Development (2021–2030) as it relates directly to the ten-year trajectory that we envision in terms of our aim of not only analysing but also effecting Just Transitions in place; specifically, in relation to sustainability, innovation, and inclusion in emerging Blue Economies. As we have seen in this volume, the question of people and stakeholders within Blue Economies, the place of local and regional dynamics and geographies, and the shift to just transitions all play an instrumental role in not only the understanding of the Blue Economy (BE) but also in the facilitation of a sustainable implementation. The BE concept is seen as a potential driver of low-carbon, ecologically sustainable coastal and marine economies in policy debates (Hadjimichael, 2018). The attraction is obvious. BE champions have highlighted the potential for low-impact food production, support of peripheral economies, innovation in materials production, and the vast scope for renewable energy generation and therefore see the BE as a promising "cog" in a future low-carbon society. However, differing definitions of the BE concept have led to numerous conceptualisations in which the original socio-ecological perspective focused on human well-being, equity, and justice is replaced with business-as-usual planning centred on ocean-based economic growth (Bennett et al., 2019). Questions regarding how to ensure and activate sustainability, ocean health, justice, and equity in established and emerging BE industries materialise as a result of the often-competing understandings surrounding the BE paradigm. Three commonalities exist across the different interpretations of the BE, through commodification (valuation of natural capital), delimitation (including a special importance given to Marine Spatial Planning), and securitisation (with maritime

DOI: 10.4324/9781003280248-20

security as both an enabler and a sector of the BE). These common features are not without risks, as they can be associated with blue grabbing, dispossession of traditional users, and territorial and natural resources disputes (Voyer et al., 2018). It has been argued that the wide endorsement for the BE actually proceeds from existing dominant discourses (including on development and sustainability), which results in a lack of innovative thinking and the spread of potentially harmful practices at sea (Germond-Duret, 2022). Hence, whereas "sustainability" and justice are supposed to be inherently embedded in the concept of BE, it is not necessarily evident. There is a risk that the BE translates into unsustainable and unjust exploitation of coastal and marine resources. There is also a risk that people affected by BE initiatives are not adequately represented in policy development, for example, as argued by Barbesgaard (2018) in the context of sustainable fisheries and blue carbon.

The appropriate design and planning of BE initiatives and projects is therefore important for delivery of triple-bottom-line (economy, society, ecology) sustainability goals for coastal communities. The transitions literature provides some pointers.[1] Policy shifts to longer term perspectives and approaches are critical for environmental sustainability, as argued by transitions scholars such as Geels (2011) and Kemp et al. (2007). Indeed, the ever-expanding literature on socio-technical transitions has provided important insights into the systems perspectives of large-scale transformations to sustainability, such as typified by the emergence of a sustainable BE (Farla et al., 2012). Köhler et al. (2017) describe the characteristics of transition approaches which have contributed significantly to debates on sustainability: a concern with the multi-dimensionality of transitions, entailing multiple interacting elements such as markets, cultural meanings, infrastructures, policies, industry structures, etc.; a recognition of multi-actor processes, whereby transitions are enacted by a range of actors and social groups from academia, politics, industry, and civil society; tensions between impulses for radical change and path dependence, inertia, and system stability; a recognition of normative dimensions, values, and the importance of the political dimension, and finally, a recognition of the central role to be played by public policy in shaping the nature of transitions. Markard (2017) argues that these features provide an effective means to study the "grand sustainability challenges."

Applying these concepts to BE, it is clear that socio-technical transition processes are highly dependent on the geographical context and the multi-scale relationships and interactions prevalent in socio-economic and socio-political processes (Morrissey & Heidkamp, 2019). The local context, and importantly, local stakeholders, become central to transition processes. Unfortunately, the transition discourse to date has been characterised by an aspatiality or by implicit assumptions about spatial homogeneity (Bridge et al., 2013; Morrissey & Heidkamp, 2019). Transitions, however, are geographical processes, which happen in situated places (Hansen & Coenen, 2015). A consideration of space is important so that unevenness, heterogeneity, and asymmetry in socio-technical systems can be acknowledged and critically understood (Raven et al., 2012). The potential for sustainability transitions will differ across regions (Hansen & Coenen, 2015), influenced by existing uneven development and potentially leading to further differentiation in spatial patterns of development (Balta-Ozkan et al., 2015; Bridge et al., 2013). In addition to this, as coastal-zone sustainability policies are socially constructed, the alignment and co-operation of stakeholders with divergent interests and priorities are required to deliver sustainability in practice (Davos, 1998; Morrissey & Heidkamp, 2019). Bennett et al. (2019) and Garland et al. (2019) have already pointed out the importance of focusing on a transition towards a just and sustainable BE; however, what matters now is that we do not continue to highlight the contradictions and potential problems but instead imagine a way forward that offers practical solutions. Essentially, a just transition in the coastal zone and the development of sustainable Blue Economies is dependent on assuring distributional, recognitional, and procedural justice (Bennett et al., 2021). In other words, blue justice refers to the fairness of the distribution of benefits and burdens; the recognition of the diversity of views, values, knowledge, rights, etc. of people involved; and the inclusiveness of the decision-making process. Assuring a just allocation of resources, fairness in decision-making, and a positive regard for social difference and the centrality of socially democratic processes must be made clear in the transition towards a sustainable BE. Innovative approaches are needed to maximise the possibilities of just transitions within the BE. We, the authors, propose a model based on a "top-led bottom-fed" approach, a greater visibility of the diverse economies composing the BE, and a transdisciplinary methodology.

16.2 An Innovative Approach for just Transitions within the Blue Economy

16.2.1 A "Top-led bottom-fed" Approach

Enacting a just and sustainable BE will require a decision-making process in which all relevant actors are involved in its shaping rather than a top-down approach or a pure—often uncritically advocated for—bottom-up approach. We suggest a less rigid "top-led bottom-fed" approach (Kokorsch, 2018). This approach can be defined as a balancing of expertise from the top [...] with the highest amount of local input and knowledge, as well as consideration of those who have to implement changes (Kokorsch, 2018). For some transition projects, outside expertise and guidance is needed in a well-balanced and reciprocal setting. Legitimacy increases when decisions and strategies are arrived at mutually between different actors: the "top" has to act respectfully and accept input from the "bottom." In turn, the "bottom" has to accept restrictions and limitations. Attempts to implement such an inclusive approach are reflected in adaptive co-management and the ideas of devolution, delegation, and decentralisation in the decision-making process (Chambers & Kokorsch, 2017). Adaptive co-management and the suggested "top-led bottom-fed" approach further share four compatible elements, namely, learning-by-doing, multiple knowledge systems, flexible management structures, and advanced collaboration through power sharing (Kokorsch et al., 2015). A consequent implementation of such approaches eventually leads to changes of governance norms and protocols and a fundamental paradigm shift of the underlying system (Armitage et al., 2009; Kokorsch, 2018). Jasanoff's (2018) "humble approach" to transitions may be useful in this context, emphasising a move away from traditional top-down command and control hierarchal governance models and a renewed focus on deliberately building societal capacity for collective action. This suggests the development of nuanced governance models which consider local values and identities, aligned with a capabilities approach (Pasgaard and Dawson, 2019). In fact, stakeholder commitment is vital to achieving a comprehensive understanding of the range of needs and values attributed to the coastal and marine environment (Mason et al., 2015). Furthermore, appropriate stakeholder engagement can enable local communities to take a degree of "ownership" over local development trajectories (Barker, 2016; Morrissey & Heidkamp, 2019). A "top-led bottom-fed" approach is especially significant when considering the "hidden aspects of the Blue Economy."

16.2.2 Blue Economy = What Economies?

We propose to adopt Gibson-Graham's (2002) economic iceberg model to visualise, analyse, and understand the BE in full. What is usually regarded as the economy—wage labour, market exchange of commodities, and capitalist enterprise—comprises but a small subset of the activities by which people produce, exchange, and distribute (Figure 16.1). The visible part of the iceberg only accounts for the visible economy, while the submerged part of the iceberg accounts for the multitude of other ways people and organisations interact in "diverse economies."

Examples of "diverse economies" within the BE go beyond subsistence fishing and include aspects such as local food networks for fish in the Icelandic Westfjords exemplifying the power dynamics that represent the larger inequalities in fishery management systems designed for globalised food networks (Smith & Chambers, 2015), illegal abalone poaching in South Africa and how the history and policies of the state contribute to the ongoing poaching taking place (Lambrechts & Goga, 2016), co-operatives within small-scale fishing industries developing as a direct response to industrial fishers' growth imperative (Ertör-Akyazi, 2020), just to highlight a few examples. These instances of "diverse economies" highlight the risks associated with blue grabbing, dispossession of traditional users, and territorial and natural resource disputes (Voyer et al., 2018) which in turn translates into a risk of unsustainable and unjust exploitation of coastal and marine resources under the BE.

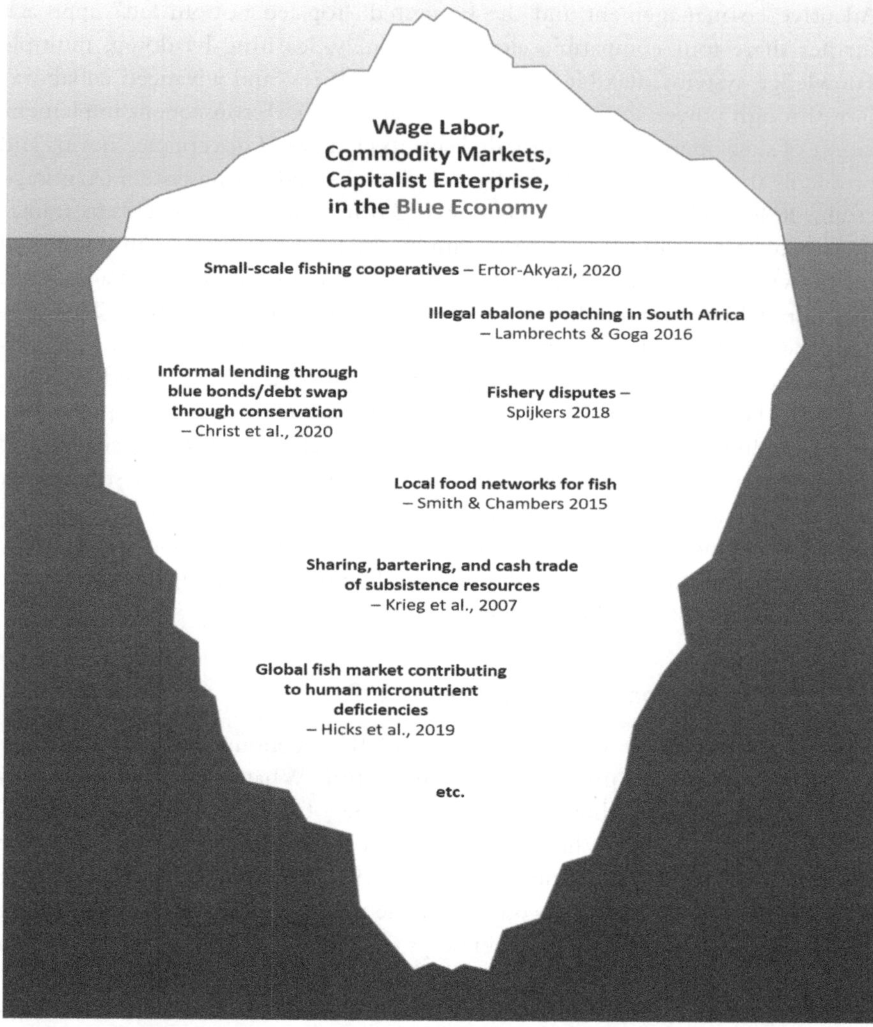

FIGURE 16.1 The Blue Economy iceberg (adapted from Diverse Economies Iceberg by Community Economies Collective which is licensed under a Creative Commons Attribution-ShareAlike 4.0 International License).

To better understand the nature of a just and sustainable BE, justice as related to social sustainability must be more deliberately considered, including the appropriate consideration of the capabilities, needs, and priorities of diverse communities. For example, there is evidence that the global fish market is contributing to human micronutrient deficiencies in Namibia's fishing communities (Hicks et al., 2019); in the Seychelles, issues with drug-use amongst its population may contribute to problems in fulfilling BE goals (Hale, 2019); tourism can add pressure to communities in Small Islands Developing States (Hampton & Jeyacheya, 2020; see also Torres Chapter 9 this volume); these examples are indicative of increasing evidence of emerging blue economies with embedded injustice features. Therefore, we aim to make "diverse economies" and justice/injustice in the context of the BE visible.

16.2.3 A Transdisciplinary and Inclusive Approach

The United Nations Decade of Ocean Science for Sustainable Development aims to "reverse the cycle of decline in ocean health and gather ocean stakeholders worldwide behind a common framework that will ensure ocean science can fully support countries in creating improved conditions for sustainable development of the Ocean" (UNESCO, 2021). We argue that the ambition of creating such a common framework requires the adherence to core principles of transdisciplinarity: (1) a problem-oriented research approach (with a normative focus on sustainability transitions), (2) a multidisciplinary framework of analysis, (3) the involvement of non-scientific actors in the process of knowledge creation, and (4) an understanding that sustainability challenges are highly context-specific (Dubielzig & Schaltegger, 2004; Padmanabhan, 2017; Renn, 2021). This is best realised through the creation of a transdisciplinary framework of analysis based on a quadruple helix approach to stakeholder engagement—involving government, academia, the business community, and civil society, built around "local-level collaboration and innovation in ... planning and development" (Hampton & Jeyacheya, 2020). Further, an inclusive Transdisciplinary Action Research (TAR) methodology represents a promising avenue in which the diversity of such stakeholder groups can mutually engage to prepare for wider earth system change, e.g., accelerated climate change (Burch et al., 2019) and envision and co-create BE futures. TAR allows for the active reimagination of a BE that is developed and driven by co-learning and shared experiences, where actors from different disciplines and spheres of experience gain a better mutual understanding and find commonality through an open door of collaboration (Heidkamp et al., 2021) and are thus able to generate context-specific knowledge and guidance for action.

16.3 Conclusion

The message we would like to convey is a message of inclusion: inclusion of people, of economies, of disciplines. We are facing a crucial ten-year period for sustainability, with coastal communities on the "front-line" of climate response and adaption. An innovative top-led bottom-fed approach that enables multi-stakeholder activation and empowerment, recognises the diversity of economies, and facilitates energised transdisciplinary dialogues has the potential to unravel the complexity of the BE and work towards its just and sustainable implementation. In the short term, the post-COVID-19 recovery represents a once-in-a-generation opportunity to radically reframe coastal and marine economies (see Morrissey & Heidkamp, 2022). The timing has never been more favourable, or indeed more critical, to urgently design, plan, and deliver sustainable, innovative, and just Blue Economies.

Note

1 Lachman (2013) provides a detailed review of transition approaches.

References

Armitage, D. R., Plummer, R., Berkes, F., Arthur, R. I., Charles, A. T., Davidson-Hunt, I. J., … & Wollenberg, E. K. (2009). Adaptive co-management for social–ecological complexity. *Frontiers in Ecology and the Environment*, 7(2), 95–102.

Barbesgaard, M. (2018). Blue growth: savior or ocean grabbing? *The Journal of Peasant Studies*, 45 (1), 130–149.

Balta-Ozkan, N., Watson, T., & Mocca, E. (2015). Spatially uneven development and low carbon transitions: Insights from urban and regional planning. *Energy Policy*, 85, 500–510. https://doi.org/10.1016/j.enpol.2015.05.013

Barker, P. (2016). *UK offshore wind—Dodging the cuts*. Retrieved June 2, 2016, from http://www.maritimejournal.com/news101/marine-renewable-energy/uk-offshore-wind-dodging-the-cuts2#sthash.mcO3NvKl.dpuf

Bennett, N. J., Cisneros-Montemayor, A. M., Blythe, J., Silver, J. J., Singh, G., Andrews, N., … & Sumaila, U. R. (2019). Towards a sustainable and equitable Blue Economy. *Nature Sustainability*, 2(11), 991–993.

Bennett, N. J., Blythe, J., White, C. S., & Campero, C. (2021). Blue growth and blue justice: Ten risks and solutions for the ocean economy. *Marine Policy*, 125, 104387.

Bridge, G., Bouzarovski, S., Bradshaw, M., & Eyre, N. (2013). Geographies of energy transition: Space, place and the low-carbon economy. *Energy Policy*, 53, 331–340. https://doi.org/10.1016/j.enpol.2012.10.066

Burch, S., Gupta, A., Inoue, C. Y., Kalfagianni, A., Persson, Å., Gerlak, A. K., … & Zondervan, R. (2019). New directions in earth system governance research. *Earth System Governance*, 1, 100006.

Chambers, C., & Kokorsch, M. (2017). The social dimension in Icelandic fisheries governance. *Coastal Management*, 45(4), 330–337.

Davos, C. A. (1998). Sustaining co-operation for coastal sustainability. *Journal of Environmental Management*, 52(4), 379–387.

Dubielzig, F., & Schaltegger, S. (2004). *Methoden transdisziplinärer Forschung und Lehre: ein zusammenfassender Überblick*. Universität Lüneburg Lehrstuhl f. Betriebswirtschaftslehre.

Ertör-Akyazi, P. (2020). Contesting growth in marine capture fisheries: The case of small-scale fishing cooperatives in Istanbul. *Sustainability Science*, 15(1), 45–62.

Farla, J., Markard, J., Raven, R., & Coenen, L. (2012). Sustainability transitions in the making: A closer look at actors, strategies and resources. *Technological Forecasting and Social Change*, 79(6), 991–998. http://doi.org/10.1016/j.techfore.2012.02.001

Garland, M., Axon, S., Graziano, M., Morrissey, J., & Heidkamp, C. P. (2019). The Blue Economy: Identifying geographic concepts and sensitivities. *Geography Compass*, 13(7), e12445.

Geels, F. W. (2011). The multi-level perspective on sustainability transitions: Responses to seven criticisms. *Environmental Innovation and Societal Transitions*, 1(1), 24–40. http://doi.org/10.1016/j.eist.2011.02.002

Germond-Duret, C. (2022). Framing the Blue Economy: Placelessness, development and sustainability. *Development and Change*, 53(2), 308–334.
Gibson-Graham, J. K. (2002). A diverse economy: Rethinking economy and economic representation. *Consultado a*, 19, 2017.
Hale, K. (2019, December 11). Seychelles blue bond economy threatened by heroin epidemic. *Forbes*. Seychelles Blue Bond Economy Threatened by Heroin Epidemic (forbes.com).
Hadjimichael, M. (2018). A call for a blue degrowth: Unravelling the European Union's fisheries and maritime policies. *Marine Policy*, 94, 158–164.
Hampton, M. P., & Jeyacheya, J. (2020). Tourism-dependent small islands, inclusive growth, and the Blue Economy. *One Earth*, 2(1), 8–10.
Hansen, T., & Coenen, L. (2015). The geography of sustainability transitions: Review, synthesis and reflections on an emergent research field. *Environmental Innovation and Societal Transitions*, 17, 92–109. https://doi.org/10.1016/j.eist.2014.11.001
Heidkamp, C. P., Garland, M., & Krak, L. (2021). Enacting a just and sustainable Blue Economy through transdisciplinary action research. *The Geographical Journal*. https://doi.org/10.1111/geoj.12410
Hicks, C. C., Cohen, P. J., Graham, N. A., Nash, K. L., Allison, E. H., D'Lima, C., … & MacNeil, M. A. (2019). Harnessing global fisheries to tackle micronutrient deficiencies. *Nature*, 574(7776), 95–98.
Jasanoff, S. (2018). Just transitions: A humble approach to global energy futures. *Energy Research & Social Science*, 35, 11–14. https://doi.org/10.1016/j.erss.2017.11.025
Kemp, R., Rotmans, J., & Loorbach, D. (2007). Assessing the Dutch energy transition policy: How does it deal with dilemmas of managing transitions? *Journal of Environmental Policy & Planning*, 9(3–4), 315–331. http://doi.org/10.1080/15239080701622816
Köhler, J., Geels, F., Kern, F., Onsongo, E., & Wieczorek, A. (2017). *A research agenda for the Sustainability Transitions Research Network*. Manchester: Sustainability Transitions Research Network (STRN), Sustainable Consumption Institute, University of Manchester.
Kokorsch, M. (2018). Mapping resilience—Coastal communities in Iceland. *Ph.D. dissertation*. Reykjavík: University of Iceland.
Kokorsch, M., Karlsdóttir, A., & Benediktsson, K. (2015). Improving or overturning the ITQ system? Views of stakeholders in Icelandic fisheries. *Maritime Studies* 14(1): 1–22.
Kokorsch, M., & Benediktsson, K. (2018). Where have all the people gone? The limits of resilience in coastal communities. *Norsk Geografisk Tidsskrift-Norwegian Journal of Geography*, 72(2), 97–114.
Lachman, D. A. (2013). A survey and review of approaches to study transitions. *Energy Policy*, 58, 269–276.
Lambrechts, D., & Goga, K. (2016). Money and marginalisation: The lost war against Abalone poaching in South Africa. *Politikon*, 43(2), 231–249.
Markard, J. (2017). Sustainability transitions: Exploring the emerging research field and its contribution to management studies, in: *Proceedings of the 33rd EGOS colloquium*, Copenhagen.
Mason, C. M., Lim-Camacho, L., Scheepers, K., & Parr, J. M. (2015). Testing the water: Understanding stakeholder readiness for strategic coastal and marine management. *Ocean & Coastal Management*, 104, 45–56.
Morrissey, J. E., Heidkamp, C. P. (2019). A transitions perspective on coastal sustainability, in: Heidkamp, C. P., Morrissey, J. E., (Eds.), *Towards coastal resilience and sustainability*. Abingdon: Routledge, pp. 13–32.
Morrissey, J., & Heidkamp, C. P. (2022). Sustainability after COVID: Pillars for a just transition. *Environmental Sustainability*. Jun 8, 1–9.

Padmanabhan, M. (Ed.). (2017). *Transdisciplinary research and sustainability: Collaboration, innovation and transformation*. Abingdon: Routledge.

Pasgaard, M., & Dawson, N. (2019). Looking beyond justice as universal basic needs is essential to progress towards 'safe and just operating spaces.' *Earth System Governance* 2, 100030. https://doi.org/10.1016/j.esg.2019.100030

Raven, R., Schot, J., & Berkhout, F. (2012). Space and scale in socio-technical transitions. *Environmental Innovation and Societal Transitions*, 4, 63–78. https://doi.org/10.1016/j.eist.2012.08.001

Renn, O. (2021). Transdisciplinarity: Synthesis towards a modular approach. *Futures*, 130, 102744.

Smith, J. G., & Chambers, C. P. (2015). Where are all the fish? Local fish networks in the Westfjords of Iceland. *Environment, Space, Place*, 7(2), 15–40.

UNESCO. (2021). *United Nations decade of ocean science for sustainable development (2021–2030)*. https://en.unesco.org/ocean-decade [accessed 10-20-2021].

Voyer, M., Quirk, G., McIlgorm, A., & Azmi, K. (2018). Shades of blue: What do competing interpretations of the Blue Economy mean for oceans governance? *Journal of Environmental Policy & Planning*, 20(5), 595–616.

INDEX

Italicized and **bold** pages refer to figures and tables respectively, and page numbers followed by "n" refer to notes.